Schlumberger Geoelectric Sounding in Ground Water

(Principles, Interpretation and Application)

HARI PADA PATRA
and
SANKAR KUMAR NATH

Department of Geology & Geophysics
Indian Institute of Technology
Kharagpur, West Bengal, India

A.A. BALKEMA / ROTTERDAM / BROOKFIELD / 1999

ISBN 90 5410 789 8

A.A. Balkema, P.O. Box 1675, 3000 BR Rotterdam, Netherlands
Fax: +31.10.4135947; E-mail: balkema@balkema.nl
Internet site: http://www.balkema.nl

Distributed in USA and Canada by
A.A. Balkema Publishers, Old Post Road, Brookfield, VT 05036-9704, USA
Fax: 802.276.3837; E-mail: Info@ashgate.com

Preface

The book entitled "Direct Current Geoelectric Sounding—Principles and Interpretation" was published through Elsevier Publishing Company, Amsterdam earlier in 1968 by the senior author of the present volume. The earlier volume was based on lectures, laboratory exercises, and field problems on geoelectric sounding, organized for senior students in exploration geophysics at the Indian Institute of Technology, Kharagpur, West Bengal (India). Interpretation method detailed in the book was the curve matching technique, avoiding computers as far as practicable. Curve matching technique based on available published master curves (Mooney and Wetzel, 1956; Compagnie Generale de Geophysique, 1955; E.A.E.G, 1963 and 1969; for example) now forms the basis of preliminary interpretation in the indirect approach. These interpreted layer parameters are at present, used as "initial guess" in the computer-aided direct approach.

Developments in computer software over the years have compelled the geophysicists to use the direct approach through inversion of resistivity data, giving final layer parameter values as output with digitized field apparent resistivity given as input. Accordingly, computer programs of the forward and inversion problems in geoelectrics are incorporated in the present volume, in detail. The book, thus, gives the students of exploration geophysics a comprehensive idea and elaborate techniques of interpretation of Schlumberger Vertical Electrical Sounding (VES) curves, with a reasonable accuracy, a minimum number of charts and an adequate number of computer softwares for the purpose.

Although the book is meant for students, it can be profitably used by professional exploration geologists and geophysicists engaged in Schlumberger vertical electrical sounding for exploration of ground water.

Civil engineers, particularly those interested in environmental problems, evaluating shallow subsurface hazard zones and bed rock depth, will find this book a self-sufficient and useful aid in their profession. Agricultural scientists and engineers interested in the exploration of ground water for irrigation and other projects may also find this book of value.

It is a pleasure to thank the Director, Prof. Amitabha Ghosh for his kind permission for publication of the book and Prof. H.C. Dasgupta, Head of the Department of Geology & Geophysics for his good wishes and encouragement in preparing the manuscript.

The cooperation extended by the members of the families of the authors is highly appreciated without whose sacrifices the manuscript could not have been completed in time.

October 1998

H.P. PATRA
S.K. NATH

Contents

1 *Introduction*

Geoelectricity is a member of group of sciences known as the geophysical sciences. It deals with the electrical state of the earth and includes discussions on the electrical properties of rocks and minerals under different geological environments, as well as their influences upon various geophysical phenomena. Geoelectric exploration (or, more simply, electrical exploration) is a major branch of exploration geophysics. It uses the principle of geoelectricity for geological mapping of concealed structures, for the exploration and prospecting of ores, minerals and oil, and in the solution of engineering geological problems.

It was formerly believed that geoelectric methods were suitable only for shallow exploration, and such methods were, therefore, primarily used for mining and engineering geophysical problems. Today, however, modern developments and refined techniques of interpretation have considerably increased the depth of investigation. It is now established (particularly through dipole electrical sounding) that exploration by geoelectric methods may be carried out with reasonable accuracy down to depths of 8-10 km.

Geoelectric exploration consists of exceedingly diverse principles and techniques, and utilizes both stationary and variable currents produced either artificially or by natural processes. One of the most widely used method of geoelectric exploration is known as the resistivity method. In this, a current (a direct or very low frequency alternating current) is introduced into the ground by two or more electrodes, and the potential difference is measured between two points (probes) suitably chosen with respect to the current electrodes. The potential difference for unit current sent through the ground is a measure of the electrical resistance of the ground between the probes. The resistance is a function of the geometrical configuration of the electrodes and the electrical parameters of the ground.

Broadly speaking, we can distinguish two types of resistivity measurements. In the first, known as geoelectric profiling or mapping, the electrodes and probes are shifted along predetermined lines without changing their relative configurations. This gives us an idea of the surface variation of resistance values within a certain depth and also along lateral directions. In the second method, known as geoelectric sounding, the positions of the electrodes are changed with respect to a fixed central point (known as the sounding point). In this way, the measured resistance values at the surface reflect the vertical distribution of resistivity values in a geological section (horizontal discontinuities). In this book we are primarily concerned with resistivity sounding only.

The two types of electrode configurations which are most frequently used in resistivity sounding are known as: (i) symmetrical and (ii) dipole. While dipole arrangement is used for deeper investigations of the order of kilometre, symmetrical arrangements (Wenner and Schlumberger) are used in shallower hydrogeological investigations. Out of the symmetrical arrays, Schlumberger is universally used, nowadays, particularly, in ground water exploration. In the present volume, Schlumberger electrode arrangement is exclusively dealt with although the elements of Wenner and dipole arrangements are very briefly mentioned at the outset. Conventional and computer-aided latest interpretation methods for Schlumberger sounding curves are detailed in this book showing the utility in the present-day ground water surveys.

The main text is divided into four chapters, which are briefly described below:

Chapter 2 deals with the theoretical foundations of Schlumberger geoelectric sounding, since a sound understanding of basic principles is essential for students specializing in this field. This chapter helps to create a solid background for what will be discussed in subsequent chapters. A reader lacking the mathematical background necessary for a grasp of these concepts may skip over this chapter and still follow, without difficulty, the interpretation procedures explained in the chapters following.

Chapter 3 gives a detailed account of the background for preliminary interpretation of Schlumberger geosounding field curves. This includes a thorough description of the approach using curves and charts reproduced within the book, comprising theoretical two-layer master curves and Ebert charts. The interpreted layer parameters serve as input for the forward method for comparison or as initial guess input data for inversion of resistivity data.

Chapter 4 outlines the method of resistivity inversion in the final computer-aided interpretation of Schlumberger sounding field curves, giving the modified accurate values of the layer parameters. Several useful inversion methods are detailed leading to a quantitative interpretation with programme listings given in the appendices. Performances of some methods are compared with regard to their utility.

Chapter 5 discusses some field applications of Schlumberger vertical electrical soundings for ground water, in practice.

2 *Theoretical Background*

This chapter deals with the theory of current flow in a horizontally stratified earth. A proper understanding of the theory is needed for the appreciation of various interpretation techniques dealt with in later chapters. Some basic concepts regarding anisotropy and apparent resistivity are introduced followed by an approximate method of computation of apparent resistivity.

2.1 Current Flow in a Homogeneous Earth

The flow of current in a medium is based on the principle of conservation of charge and is expressed by the relation:

$$div \bar{J} = -\frac{\partial \rho'}{\partial t} \qquad (2.1)$$

where $\bar{J}$ is the current density (A/m^2) and ρ' is the charge density (C/m^3). This relation (2.1) is also known as the 'equation of continuity'. For stationary current (2.1) reduces to:

$$div \bar{J} = 0 \qquad (2.2)$$

If ρ is the resistivity (ohm-m) of the medium, then the current density $\bar{J}$ is related to the electric field intensity E (V/m) by means of Ohm's law, which is given as:

$$\bar{J} = \frac{1}{\rho} \bar{E} = -\frac{1}{\rho} \text{grad } V \qquad (2.3)$$

where **V** is the electric potential (volts). For an isotropic medium, ρ is a scalar function of the point of observation, and $\bar{J}$ is in the same direction as $\bar{E}$. In an anisotropic medium, however, $\bar{J}$ has a directive property and, in general, is not in the direction of $\bar{E}$. This calls for a modified

Ohm's law in an anisotropic medium where conductivity forms a symmetric tensor having six components. The current flow in an anisotropic medium and other aspects will be treated later in this chapter. For an isotropic medium we get from relations (2.2) and (2.3):

$$div\left(\frac{1}{\rho}\,\text{grad}V\right)=0 \tag{2.4}$$

or:

$$\text{grad}\left(\frac{1}{\rho}\right)\cdot\text{grad }V+\frac{1}{\rho}\,div\text{ grad }V=0 \tag{2.5}$$

This is the fundamental equation of electrical prospecting with direct current. If the medium is homogeneous, ρ is independent of the coordinate axes and hence:

$$\nabla^2 V=0 \tag{2.6}$$

This is Laplace's equation which can also be derived from Maxwell's equations (I) and (IV) given as follows:
Maxwell's equations:

$$\nabla\times\vec{E}+\frac{\partial\vec{B}}{\partial t}=0 \qquad \text{(Faraday's law)} \tag{I}$$

$$\nabla\times\vec{H}-\frac{\partial\vec{D}}{\partial t}=\vec{J} \qquad \text{(Ampere's law)} \tag{II}$$

$$\nabla\cdot\vec{B}=0 \qquad \text{(solenoidal } \vec{B}) \tag{III}$$

$$\nabla\cdot\vec{D}=\rho' \qquad \text{(Coulomb's law)} \tag{IV}$$

These equations are known as Maxwell's equations because it was Maxwell who assembled these laws of physics in a differential form, found the anomaly and introduced the displacement current density for a generalized representation of electromagnetic fields. In the rationalized MKS system followed in this book the units are:

$\vec{B}$ = magnetic induction in weber/square metre (Wb/m²)

$\vec{H}$ = magnetic field intensity in ampere/metre (A/m)

$\vec{E}$ = electric field intensity in volts/metre (V/m)

$\vec{D}$ = electric displacement in coulombs/square metre (C/m²)

$\vec{J}$ = electric current density in ampere/square metre (A/m²)

ρ' = electric charge density in coulombs/cubic metre (C/m³)

From analogy with $\vec{J}$, $\partial \vec{D}/\partial t$ is known as displacement current density.

Laplace's equation may be thought of as a special case, or to be exact a direct consequence of Maxwell's equations I and IV. Eq. I which reads as:

$$\nabla \times \vec{E} + \frac{\partial \vec{B}}{\partial t} = 0 \qquad (2.7)$$

reduces in the static or stationary case with $\partial \vec{B}/\partial t = 0$ to the form:

$$\nabla \times \vec{E} = 0 \qquad (2.8)$$

Eq. (2.8) means that the line integral of the electric fields intensity $\vec{E}$ around any closed path is zero and therefore the field is conservative, a necessary and sufficient condition of the existence of a scalar potential (V) whose gradient is $\vec{E}$, i.e.:

$$\vec{E} = - \nabla V \qquad (2.9)$$

From eq. IV, then we have

$$\nabla \cdot \vec{E} = \rho'/\varepsilon \qquad (2.10)$$

or:

$$\nabla \cdot \nabla V \equiv \nabla^2 V = - \rho'/\varepsilon \quad \text{(Poisson's equation)} \qquad (2.11)$$

At points free of charge ($\rho' = 0$) we obtain:

$$\nabla^2 V = 0 \quad \text{(Laplace's equation)} \qquad (2.12)$$

Before proceeding to discussions on current flow within earth we shall write Laplace's eq. (2.12) in cylindrical and spherical polar coordinates. However, it is easier to remember Laplace's equation in generalized orthogonal curvilinear coordinates from which we can derive other coordinate systems by substitution in the manner shown below.

In orthogonal curvilinear coordinates we have:

$$\nabla^2 V = \frac{1}{h_1 h_2 h_3} \left[\frac{\partial}{\partial u_1} \left(\frac{h_2 h_3}{h_1} \frac{\partial V}{\partial u_1} \right) + \frac{\partial}{\partial u_2} \left(\frac{h_1 h_3}{h_2} \frac{\partial V}{\partial u_2} \right) + \frac{\partial}{\partial u_3} \left(\frac{h_1 h_2}{h_3} \frac{\partial V}{\partial u_3} \right) \right]$$

$$(2.13)$$

Here the values of the parameters in different systems are as follows:

(i) Rectangular coordinates:
$$u_1 = x, \quad u_2 = y, \quad u_3 = z$$
$$h_1 = 1, \quad h_2 = 1, \quad h_3 = 1$$

(ii) Cylindrical polar coordinates:
$$u_1 = r, \quad u_2 = \varphi, \quad u_3 = z$$
$$h_1 = 1, \quad h_2 = r, \quad h_3 = 1$$

(iii) Spherical polar coordinates:
$$u_1 = r, \quad u_2 = \theta, \quad u_3 = \varphi$$
$$h_1 = 1, \quad h_2 = r, \quad h_3 = r \sin \theta$$

then we have in a cylindrical system:

$$\nabla^2 V = \frac{1}{r} \left[\frac{\partial}{\partial r} \left(r \frac{\partial V}{\partial r} \right) + \frac{\partial}{\partial \varphi} \left(\frac{1}{r} \frac{\partial V}{\partial \varphi} \right) + \frac{\partial}{\partial z} \left(r \frac{\partial V}{\partial z} \right) \right] = 0$$

or:

$$\frac{1}{r} \frac{\partial}{\partial r} \left(r \frac{\partial V}{\partial r} \right) + \frac{1}{r^2} \frac{\partial^2 V}{\partial \varphi^2} + \frac{\partial^2 V}{\partial z^2} = 0 \tag{2.14}$$

and in spherical polar coordinates:

$$\frac{1}{r^2 \sin \theta} \left[\frac{\partial}{\partial r} \left(r^2 \sin \theta \frac{\partial V}{\partial r} \right) + \frac{\partial}{\partial \theta} \left(\sin \theta \frac{\partial V}{\partial \theta} \right) + \frac{\partial}{\partial \varphi} \left(\frac{1}{\sin \theta} \frac{\partial V}{\partial \varphi} \right) \right] = 0$$

or:

$$\frac{1}{r^2} \left[\frac{\partial}{\partial r} \left(r^2 \frac{\partial V}{\partial r} \right) + \frac{1}{\sin \theta} \frac{\partial}{\partial \theta} \left(\sin \theta \frac{\partial V}{\partial \theta} \right) + \frac{1}{\sin^2 \theta} \frac{\partial^2 V}{\partial \varphi^2} \right] = 0 \tag{2.15}$$

Eqs. (2.14) and (2.15) are very important from the viewpoint of dc resistivity sounding and profiling on the surface of the earth in order to get subsurface information.

As a prerequisite to evaluation of potential distribution in a layered earth we must first calculate the normal potential at the surface due to a point source of current I. We find out the potential at any point (P) in an infinite homogeneous medium of resistivity ρ. Laplace's eq. (2.12) in spherical polar coordinates (2.15) with symmetry with respect to θ and φ reduces to:

$$\frac{\partial}{\partial r} \left(r^2 \frac{\partial V}{\partial r} \right) = 0 \tag{2.16}$$

On integration we get:

$$V = C_1 + \frac{C_2}{r}$$ (2.17)

As the potential is taken to be zero at a large distance from the source, the integration constant $C_1 = 0$. It is clear that the equipotential surfaces are spherical, and the electric field lines as well as the current lines are radial. The current density at a distance r may be written as:

$$J = -\frac{1}{\rho}\frac{\partial V}{\partial r} = \frac{1}{\rho}\frac{C_2}{r^2}$$

Thus, the total current flowing out of a spherical surface of radius r is:

$$4\pi r^2 J = \frac{4\pi}{\rho}C_2$$

Since this is equal to I, the total current introduced at P, the constant C_2 is given by:

$$C_2 = I\rho/4\pi$$

For a semi-infinite medium, i.e., when the current is introduced into a homogeneous ground, the total current flowing out of a hemispherical surface of radius r is given by the relation, $2\pi r^2 J = (2\pi/\rho)C_2$, and the constant C_2 is equal to $I\rho/2\pi$.

Thus, the potential at any point due to a current source at the surface of a homogeneous earth is:

$$V = \frac{I\rho}{2\pi}\frac{1}{r}$$ (2.18)

In practice, the current is introduced into the ground by means of two electrodes, i.e., a source and a sink; and the potential at any point, from the "principle of superposition", due to this "bipolar" arrangement is:

$$V = \frac{I\rho}{2\pi}\left(\frac{1}{r_1} - \frac{1}{r_2}\right)$$ (2.19)

where r_1 and r_2 are the distances of the point P from the source and the sink, respectively.

2.2 Resistivity Measurement

Consider that a direct current of strength I is introduced into a homogeneous and isotropic earth by means of two point electrodes A and B (Fig. 2.1). The potential difference between the two points M and N on the surface is given by using eq.(2.19)

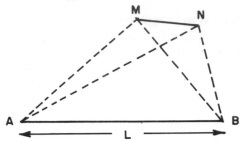

Fig. 2.1 Point electrodes over a homogeneous and isotropic earth. A, B = point source and sink; M, N = observation points on the surface of the earth.

$$V = \frac{I\rho}{2\pi}\left[\left(\frac{1}{AM} - \frac{1}{BM}\right) - \left(\frac{1}{AN} - \frac{1}{BN}\right)\right] \tag{2.20}$$

where ρ is the resistivity of the ground. Thus, the resistivity of the homogeneous earth can be determined from the measurements on the surface.

Various electrode arrangements for A, B, M, and N have been suggested for the purpose. The ones more commonly used for resistivity sounding are: (1) symmetrical arrangement, and (2) dipole arrangement.

In the symmetrical arrangement, the points A, M, N, B are taken on a straight line such that the points M and N are symmetrically placed about the centre O of the "spread" AB (Fig. 2.2).

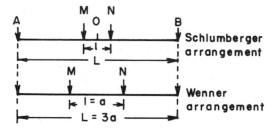

Fig. 2.2 Symmetrical electrode arrangements. Top: Schlumberger arrangement, Bottom: Wenner arrangement.

Here :

$$\Delta V = \frac{I\rho}{2\pi}\left(\frac{4}{L-l} - \frac{4}{L+l}\right) \tag{2.21}$$

which gives:

$$\rho = \frac{\pi}{4} \frac{(L^2 - l^2)}{l} \frac{\Delta V}{I} \qquad (2.22)$$

In the 'Wenner arrangement', L is taken equal to $3l$ (l is conventionally denoted by "a" in the Wenner configuration and is known as spacing or separation of the electrodes), and the resistivity is given by:

$$\rho = 2\pi a \frac{\Delta V}{I} \qquad (2.23)$$

If $L \geq 5l$, we can put $(L^2 - l^2)$ in eq. (2.22) equal to L^2 with an error less than 4%. This is known as the 'Schlumberger arrangement' (Fig. 2.2). In this case, the resistivity is given by:

$$\rho = \frac{\pi L^2}{4} \frac{\Delta V}{l} \frac{1}{I} = \frac{\pi L^2}{4} \frac{E}{I} \qquad (2.24)$$

where $E = \Delta V/l$ is (approximately) the electric intensity at the central point O. Hence, this arrangement is sometimes known as the "gradient arrangement", and Wenner's arrangement as a "potential arrangement".

The general dipole arrangement is shown in Fig. 2.3, where r is usually taken to be much larger than AB. The potential at O due to AB is given by:

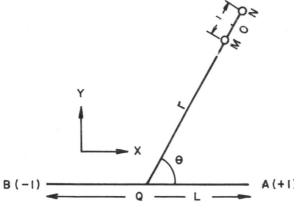

Fig. 2.3 General arrangement for dipole electrical sounding. AB = current dipole; MN = measuring dipole; Q, O = midpoints of current and measuring dipoles.

$$V = \frac{I\rho}{2\pi} \left(\frac{1}{AO} - \frac{1}{BO} \right)$$

which can be expressed in a series, and the potential may be written as:

$$V \approx \frac{I\rho L \cos\theta}{2\pi r^2} \quad \text{(for } r \gg L\text{)} \tag{2.25}$$

If r is greater than $3L$, the error of neglecting the higher-order terms is less than 3%. Thus, the potential is equal to that of a dipole of moment $I\rho L/2\pi$.

The expressions for different dipole arrangements (Fig. 2.4) may be obtained from eq. (2.25) which is outside the scope of the book. Dipole methods are used for deeper sounding for depths normally beyond 1km.

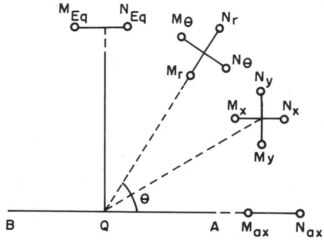

Fig. 2.4 Various arrangement for dipole sounding $M_{ax} N_{ax}$ = axial; $M_x N_x$ = parallel, $M_y N_y$ = perpendicular; $M_r N_r$ = radial; $M_\theta N_\theta$ = azimuthal; $M_{Eq} N_{Eq}$ = equatorial.

2.2.1 Apparent Resistivity

Resistivity of a material is defined as the resistance offered between the opposite faces of a unit cube. The resistivity of a semi-infinite earth can be calculated for various electrode arrangements, using the formulas given in equations 2.23, 2.24 and 2.25. For an inhomogeneous medium, on the other hand, we define a quantity $\bar{\rho}$ known as 'apparent resistivity'. The apparent resistivity of a geologic formation is equal to the true resistivity of a fictitious homogeneous and isotropic medium in which, for a given electrode arrangement and current strength I, the measured potential difference V is equal to that for the given inhomogeneous medium under consideration. The apparent resistivity depends upon the geometry and resistivities of the elements constituting the given geologic medium. Thus:

$\bar{\rho} = \bar{K}(\nabla V / I)$ where $\bar{K}$ is the geometrical coefficient having the dimension of length(m). For the different arrangements discussed in the section 2.1 on current flow in a homogeneous earth, three values of $\overline{K}$ will be given below:

(1) Symmetrical:

$$\overline{K}_w = 6.28a \qquad\qquad \text{(Wenner)}$$

$$\overline{K}_s = 0.785 \frac{(L+l)(L-l)}{l} \qquad\qquad \text{(Schlumberger)}$$

(2) Dipole:

Radial: $\qquad\qquad \overline{K}_r = \dfrac{\pi r^3}{Ll \cos\theta}$

Azimuthal: $\qquad\qquad \overline{K}_\theta = \dfrac{2\pi r^3}{Ll \sin\theta}$

Parallel: $\qquad\qquad \overline{K}_x = \dfrac{2\pi r^3}{Ll} \dfrac{1}{3\cos^2\theta - 1}$

Perpendicular: $\qquad\qquad \overline{K}_y = \dfrac{2\pi r^3}{3Ll} \dfrac{1}{\sin\theta\cos\theta}$

Equatorial: $\qquad\qquad \overline{K}_{eq} = \dfrac{2\pi r^3}{Ll}$

Axial: $\qquad\qquad \overline{K}_{ux} = \dfrac{\pi r^3}{Ll}$

2.3 Current Flow in a Homogeneous Anisotropic Earth

It has been shown that in a homogeneous anisotropic medium the conductivity forms a symmetrical tensor characterised by six components. In such a case, it is always possible to orient the coordinate axes in such a way that these axes form "principal axes of anisotropy", and the principal values of $\overline{J}$ and $\overline{E}$ are then given by (in rectangular coordinates):

$$J_x = \frac{1}{\rho_x} E_x ; \quad J_y = \frac{1}{\rho_y} E_y ; \quad J_z = \frac{1}{\rho_z} E_z$$

where x, y, z are now the principal axes. Thus, the equation of continuity (2.2) may be written in terms of the principal axes as:

$$\frac{\partial}{\partial x}\left(\frac{E_x}{\rho_x}\right) + \frac{\partial}{\partial y}\left(\frac{E_y}{\rho_y}\right) + \frac{\partial}{\partial z}\left(\frac{E_z}{\rho_z}\right) = 0$$

For a homogeneous medium this reduces to:

$$\frac{1}{\rho_x}\frac{\partial^2 V}{\partial x^2} + \frac{1}{\rho_y}\frac{\partial^2 V}{\partial y^2} + \frac{1}{\rho_z}\frac{\partial^2 V}{\partial z^2} = 0 \qquad (2.26)$$

Choose a new system of coordinates, such that:

$$\xi = x\sqrt{\rho_x}; \quad \eta = y\sqrt{\rho_y}; \quad \zeta = z\sqrt{\rho_z}$$

then eq. (2.26) reduces to Laplace's equation:

$$\frac{\partial^2 V}{\partial \xi^2} + \frac{\partial^2 V}{\partial \eta^2} + \frac{\partial^2 V}{\partial \zeta^2} = 0$$

the solution of which is:

$$V = \frac{C}{(\xi^2 + \eta^2 + \zeta^2)^{\frac{1}{2}}}$$

where C is the constant of integration. Thus, the solution to eq.(2.26) is:

$$V = \frac{C}{(\rho_x X^2 + \rho_y y^2 + \rho_z z^2)^{\frac{1}{2}}} \qquad (2.27)$$

It is seen from eq. (2.27) that the equipotential surfaces given by:

$$\rho_x x^2 + \rho_y y^2 + \rho_z z^2 = K^2$$

are ellipsoids, the axes of which coincide with the principal axes of anisotropy. The current densities are given by:

$$Cx = -\frac{1}{\rho_x}\frac{\partial V}{\partial x} = \frac{Cx}{(\rho_x x^2 + \rho_y y^2 + \rho_z z^2)^{\frac{3}{2}}}$$

$$Cy = -\frac{1}{\rho_y}\frac{\partial V}{\partial y} = \frac{Cy}{(\rho_x x^2 + \rho_y y^2 + \rho_z z^2)^{\frac{3}{2}}} \qquad (2.28)$$

$$Cz = -\frac{1}{\rho_z}\frac{\partial V}{\partial z} = \frac{Cz}{(\rho_x x^2 + \rho_y y^2 + \rho_z z^2)^{\frac{3}{2}}}$$

These equations satisfy the relation:

$$\frac{J_x}{x} = \frac{J_y}{y} = \frac{J_z}{z}$$

showing that the current lines are straight lines, spreading out radially from the source, as in the case of an isotropic medium.

The electric lines of force in an anisotropic medium form a family of curvilinear trajectories orthogonal to the equipotential surfaces. They do not coincide with the directions of the current lines except along the principal axes.

The anisotropy in a geological body may be due to several reasons. Thus, it is a well-known fact that in stratified rocks the strike offers a particularly favourable path for the flow of electric currents. The reason may be that a large number of mineral crystals possess a flat or elongated shape (mica, kaoline, etc.). At the time of their deposition, they naturally take an orientation parallel to the sedimentation. The weathered surface soil, owing to vegetable matter, growth and decay of minerals in the soil, etc., also manifests an anisotropic character. In electrical exploration, the usual practice is to characterize the electrical property of a stratified rock by two parameters, namely, the longitudinal resistivity ρ_s (parallel to the plane of stratification) and the transverse resistivity ρ_t (normal to the plane of stratification). Thus, any anisotropy in the plane of stratification is usually neglected, being very small in most practical cases.

If the plane of stratification is chosen as the xy plane, then eq. (2.26) reduces to:

$$\frac{1}{\rho_s}\left(\frac{\partial^2 V}{\partial x^2} + \frac{\partial^2 V}{\partial y^2}\right) + \frac{1}{\rho_t}\frac{\partial^2 V}{\partial z^2} = 0 \tag{2.29}$$

The equipotential surfaces are then given by:

$$x^2 + y^2 + (\rho_t/\rho_s)z^2 = \text{constant}$$

i.e., they are ellipsoids of revolution around the z-axis.

We define two more parameters of an anisotropic medium:

$$\lambda = \sqrt{\rho_t/\rho_s} \quad \text{and} \quad \rho_m = \sqrt{\rho_t\,\rho_s} \tag{2.30}$$

where λ is called the "coefficient of anisotropy" and ρ_m the "root mean square resistivity", or simply the "mean resistivity" of the medium. It is obvious from eq. (2.30) that:

$$\rho_m = \lambda\rho_s = \frac{1}{\lambda}\rho_t \tag{2.31}$$

The solution of eq. (2.29) may now be written as:

$$V = \frac{C}{\rho_s^{\frac{1}{2}} (x^2 + y^2 + \lambda^2 z^2)^{\frac{1}{2}}} \tag{2.32}$$

and the current densities are:

$$J_x = \frac{Cx}{\rho_s^{\frac{3}{2}} (x^2 + y^2 + \lambda^2 z^2)^{\frac{3}{2}}}$$

$$J_y = \frac{Cy}{\rho_s^{\frac{3}{2}} (x^2 + y^2 + \lambda^2 z^2)^{\frac{3}{2}}}$$

$$J_z = \frac{Cz}{\rho_s^{\frac{3}{2}} (x^2 + y^2 + \lambda^2 z^2)^{\frac{3}{2}}}$$

such that:

$$J = \left(J_x^2 + J_y^2 + J_z^2 \right)^{\frac{1}{2}} = \frac{C(x^2 + y^2 + z^2)^{\frac{1}{2}}}{\rho_s^{\frac{3}{2}} (x^2 + y^2 + \lambda^2 z^2)^{\frac{3}{2}}} \tag{2.33}$$

In order to find the constant of integration C, we construct around P a sphere of radius R and calculate the total current flowing out through this spherical surface.
This obviously is equal to the total current at the electrode P.
Thus:

$$I = \int_s J ds = \int_0^{2\pi}\int_0^{\pi} JR^2 \sin\theta d\theta d\varphi \tag{2.34}$$

Now: $x^2 + y^2 = R^2 \sin^2\theta$ and $z^2 = R^2 \cos^2\theta$, then eq. (2.33) becomes:

$$J = \frac{C}{\rho_s^{\frac{3}{2}} R^2 (\sin^2\theta + \lambda^2 \cos^2\theta)^{\frac{3}{2}}} = \frac{C}{\rho_s^{\frac{3}{2}} R^2 [1 + (\lambda^2 - 1)\cos^2\theta]^{\frac{3}{2}}}$$

and:

$$I = \frac{C}{\rho_s^{\frac{3}{2}}} \int_0^{2\pi} d\varphi \int_0^{\pi} \frac{\sin\theta d\theta}{[1 + (\lambda^2 - 1)\cos^2\theta]^{\frac{3}{2}}} = \frac{2\pi C}{\rho_s^{\frac{3}{2}}} \frac{2}{\lambda} = \frac{4\pi C}{\lambda \rho_s^{\frac{3}{2}}}$$

Therefore:

$$C = \frac{I}{4\pi}\lambda\rho_s^{\frac{3}{2}} \qquad (2.35)$$

and, consequently, eqs. (2.32) and (2.33) become:

$$V = \frac{I\lambda\rho_s}{4\pi(x^2 + y^2 + \lambda^2 z^2)^{\frac{1}{2}}} = \frac{I\rho_m}{4\pi R[1 + (\lambda^2 - 1)\cos^2\theta]^{\frac{1}{2}}} \qquad (2.36)$$

and:

$$J = \frac{I\lambda(x^2 + y^2 + z^2)^{\frac{1}{2}}}{4\pi(x^2 + y^2 + \lambda^2 z^2)^{\frac{3}{2}}} = \frac{I\lambda}{4\pi R^2[1 + (\lambda^2 - 1)\cos^2\theta]^{\frac{3}{2}}} \qquad (2.37)$$

For an isotropic infinite homogeneous medium: $\lambda = 1$, $\rho_m = \rho$, and: $V = \rho I/4\pi R$ and $J = I/4\pi R^2$, as given in the section on a homogeneous earth.

Let us now put the point source of current I on the surface of the ground, which is assumed to be homogeneous but is anisotropic. Assuming that air has infinite resistivity, the current density in air is zero. The value of V and J are still given by eqs. (2.32) and (2.33), but the value of C is now determined by finding the total current flow through a hemisphere of radius R, i.e., instead of eq. (2.34) we shall have:

$$I = \int_s J ds = \int_0^{2\pi}\int_0^{\frac{\pi}{2}} J R^2 \sin\theta\, d\theta\, d\varphi$$

giving:

$$C = \frac{I}{2\pi}\lambda\rho_s^{\frac{3}{2}}$$

Thus, relations (2.36) and (2.37) are replaced by:

$$V = \frac{I}{2\pi R}\frac{\rho_m}{[1 + (\lambda^2 - 1)\cos^2\theta]^{\frac{1}{2}}} = \frac{I\rho_m}{2\pi}\frac{1}{(x^2 + y^2 + \lambda^2 z^2)^{\frac{1}{2}}} \qquad (2.38)$$

and:

$$J = \frac{I}{2\pi R^2}\frac{\lambda}{[1 + (\lambda^2 - 1)\cos^2\theta]^{\frac{3}{2}}} = \frac{I\lambda}{2\pi}\frac{R}{(x^2 + y^2 + \lambda^2 z^2)^{\frac{3}{2}}} \qquad (2.39)$$

As in case of an infinite medium, here also the equipotential surfaces are ellipsoids of revolution about the axis of z, i.e., perpendicular to the plane of stratification.

In eqs. (2.38) and (2.39) we have assumed that the air-earth boundary is parallel to the plane of stratification. In order to generalize the formulas, we take two systems of coordinate, $x'y'z'$ and xyz (Fig. 2.5), in which $x'y'$ represent the air-earth boundary and xy the plane of stratification, such that the strike of the bed x is taken as x' of the new system. Let the angle of dip be α.

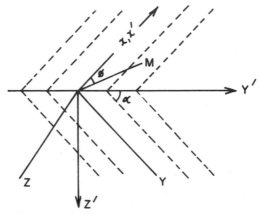

Fig. 2.5 Anisotropic half-space, xy = plane of stratification; $x'y'$ = air-earth boundary surface; $\propto$ = dip of the bed; φ = angle made by the point of observation with the strike direction.

Now:

$$x = x'$$
$$y = y' \cos \alpha + z' \sin \alpha$$
$$z = -y' \sin \alpha + z' \cos \alpha$$

Introducing the new coordinate system, and putting $z' = 0$, we get from eq. (2.38):

$$V = \frac{I\rho_m}{2\pi} \frac{1}{(x'^2 + y'^2 \cos^2 \alpha + \lambda^2 y'^2 \sin^2 \alpha)^{\frac{1}{2}}}$$

$$= \frac{I\rho_m}{2\pi} \frac{1}{[x'^2 + \{1 + (\lambda^2 - 1)\sin^2 \alpha\}y'^2]^{\frac{1}{2}}}$$

Writing $r^2 = x'^2 + y'^2$ and $\tan \varphi = \dfrac{y'}{x'}$, we get:

$$V = \frac{I \rho_m}{2\pi r} \frac{1}{[1 + (\lambda^2 - 1)\sin^2 \varphi \sin^2 \alpha]^{\frac{1}{2}}} \qquad (2.40)$$

Eq. (2.40) gives the potential at any point (M) on the surface, at a distance r from the source, and in the direction which makes an angle φ with the strike direction.

It is seen from eq. (2.40) that the equipotential lines on the surface are ellipses, with major axes oriented along the strike direction. The ratio of the semi-major to semi-minor axes are given by:

$$\frac{a}{b} = [1 + (\lambda^2 - 1)\sin^2 \alpha]^{\frac{1}{2}} \qquad (2.41)$$

which depends on the coefficient of anisotropy and the angle of dip. In the case of an isotropic medium (i.e., $\lambda = 1$), as also for horizontal beds (i.e., $\alpha = 0$), $a/b = 1$. Thus, in these cases the equipotential lines are circles about the source. For $\alpha = \pi/2$, $a = b\lambda$.

The formula (2.41) can be used to determine the coefficient of anisotropy (λ), when the dip of the beds is known, or conversely to determine the dip angle α when the anisotropy is known. The experimental procedure would be to draw the equipotential line and then determine the ratio of the axes a/b.

Differentiating eq. (2.40) with respect to r, we get the radial component of the electric field:

$$E = -\frac{\partial V}{\partial r} = \frac{I \rho_m}{2\pi r^2} \frac{1}{[1 + (\lambda^2 - 1)\sin^2 \varphi \sin^2 \alpha]^{\frac{1}{2}}} \qquad (2.42)$$

Defining apparent resistivity ($\bar{\rho}$), as in the case of an isotropic medium for a symmetrical Schlumberger spread, i.e., $\bar{\rho} = \left(\dfrac{E}{I}\right) 2\pi r^2$, we get:

$$\bar{\rho} = \frac{\rho_m}{[1 + (\lambda^2 - 1)\sin^2 \varphi \sin^2 \alpha]^{\frac{1}{2}}} \qquad (2.43)$$

It follows that, along the strike direction:

$$\varphi = 0; \quad \bar{\rho}_s = \rho_m \qquad (2.44)$$

and normal to the strike direction, $\varphi = \dfrac{\pi}{2}$:

$$\bar{\rho}_t = \frac{\rho_m}{[1 + (\lambda^2 - 1)\sin^2 \alpha]^{\frac{1}{2}}} \qquad (2.45)$$

Thus, the apparent resistivity, measured on the surface of a homogeneous anisotropic formation along the strike direction, is independent of the dip — eq. (2.44) — and is numerically equal to the mean resistivity of the formation. However, the apparent resistivity normal to the strike direction — eq. (2.45) — is dependent on the dip. Also, since the denominator of eq. (2.45) is greater than unity (except when $\alpha = 0$), it follows that except for $\alpha = 0$:

$$\bar{\rho}_t < \bar{\rho}_s \qquad (2.46)$$

For the special case $\alpha = \dfrac{\pi}{2}$:

$$\bar{\rho}_t = \frac{\rho_m}{\lambda} = \bar{\rho}_s \qquad (2.47)$$

From relation (2.46) it follows that the apparent resistivity $\bar{\rho}_t$, measured normal to the strike direction, is less than $\bar{\rho}_s$, measured along the strike direction—although it is known that the true resistivity of an anisotropic formation, normal to its stratification, is greater than that parallel to the plane of stratification. This phenomenon is called the "paradox of anisotropy".

The paradox is explained by the fact that since $\rho_s < \rho_t$, the current density along the plane of stratification is greater than along the normal to this plane.

The anisotropy studied so far in this section characterizes finely stratified rocks which appear homogeneous to the eye and, in fact, in numerous cases this corresponds to a real homogeneity in composition. This type of anisotropy is microscopic and may be called "micro-anisotropy". In electrical prospecting it is necessary to consider a second kind of anisotropy, which may be called "macro-anisotropy". In practice it is sometimes difficult to draw the boundary between the micro and the macro-anisotropy.

We may call a medium macro-anisotropic as long as the layers can be distinguished—for example, by electrical logging in a borehole. The concept of macro-anisotropy will be explained in brief in the next few paragraphs.

Macro-anisotropy results from the repetitive alternation of two different isotropic lithologic facies. When the individual layers become infinitely thin and infinitely repetitive, we obviously reach the domain of micro-anisotropy. The study of this parameter is of primary importance to the geophysicist, because the distribution of the electric field, due to two current electrodes at the surface of the ground, will be governed by the resistivity and thickness of the underlying layers in addition to the distance between the current electrodes. This effective resistivity and effective thickness are controlled by anisotropy.

In electrical prospecting, the two parameters of importance are: resistivity parallel to stratification (ρ_s), and resistivity normal to stratification (ρ_t); the physical significance of these has already been explained. Consequently, it is found that by adopting the concept of ρ_s and ρ_t for a group of layers, we are concerned with an anisotropy phenomenon; and the layers may be considered to behave as a single anisotropic layer of pseudo- or equivalent anisotropy λ. This anisotropic fictitious layer may be taken to be equivalent to another, single isotropic layer, of pseudo-resistivity ρ_e and pseudo-thickness h_e. This forms the basis of the analytic-graphical auxiliary-point method of interpretation, which will be dealt with in detail later in this book (Chapter 3).

Anisotropy plays an important role in the interpretation of layer parameters as an error is introduced in ignoring it. Surface measurements do not differentiate an isotropic bed of thickness h and resistivity ρ from anisotropic bed of thickness h/λ and resistivity ρ_m. λ being always greater than unity, this means that depth derived from isotropic concept is more than the true depth when there exists an anisotropy but is neglected. Geological control and well-logging data help in correct interpretation of such cases.

2.4 Current Flow in a Horizontally Stratified Earth

In electrical prospecting it is often necessary to determine the depth and the electrical resistivity of horizontal or nearly horizontal layers. In order to solve this problem, we should calculate the potential and the electrical field, due to a point source of current, at any point on the surface of a stratified earth.

Let us choose a cylindrical system of coordinates, with the origin at the point source A, and the z-axis vertically downward normal to the surface (Fig. 2.6). Let ρ_1, ρ_2, ...ρ_n be the resistivities, and h_1, h_2, ... h_n be the thicknesses of n layers from the top. Also let H_1, H_2, ...H_n denote the depths of the bottoms of each layer. We shall assume that the lowermost layer extends to infinity, i.e., $h_n = \infty$, $H_n = \infty$.

The Laplace equation 2.14 in cylindrical system to be satisfied at any point is (due to symmetry about φ):

$$\frac{\partial^2 V}{\partial r^2} + \frac{1}{r}\frac{\partial V}{\partial r} + \frac{\partial^2 V}{\partial z^2} = 0 \qquad (2.48)$$

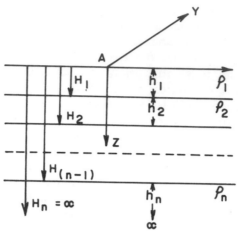

Fig. 2.6   A multi-layer earth.

the general solution of which may be written as:

$$V = \int_0^\infty [A(m)e^{-mz} + B(m)e^{mz}] J_0(mr)dm \qquad (2.49)$$

We know that the potential due to a point source of current placed on the surface of a homogeneous earth is (eq. 2.18):

$$V_0 = \frac{I\rho}{2\pi} \frac{1}{R} = \frac{I\rho}{2\pi} \frac{1}{(r^2 + z^2)^{\frac{1}{2}}}$$

where R is the distance between A and the point of observation.

The potential due to the point source at any point in the layered earth may be thought of as the sum of the potentials in a homogeneous medium and a perturbation potential, due to the boundaries, given by eq. (2.49).

Then: $V_1 = V_0 + V_1'$; $V_2 = V_0 + V_2'$; $V_i = V_0 + V_i'$; and $V_n = V_0 + V_n'$ where $V_1, V_2, ...V_n$ are the total potentials in the different strata, and V_1', $V_2', ... V_n'$ are the perturbation potentials. Perturbation potential may be physically explained as the contribution due to two series of an infinite number of images on both sides of the boundaries.

Thus, in general:

$$V_i = \frac{I\rho_1}{2\pi} \frac{1}{(r^2 + z^2)^{\frac{1}{2}}} + \int_0^\infty [A_i(m)e^{-mz} + B_i(m)e^{mz}] J_0(mr)dm \qquad (2.50)$$

The constants A_i, B_i may be determined from the respective boundary conditions. At the air-earth boundary, we must have:

$$\frac{1}{\rho_1}\frac{\partial V_1}{\partial z} = 0 \text{ at } z = 0.$$

Now:

$$V_1 = \frac{I\rho_1}{2\pi}\frac{1}{(r^2 + z^2)^{\frac{1}{2}}} + \int_0^\infty (A_1 e^{-mz} + B_1 e^{mz})J_0(mr)dm$$

Therefore:

$$\left(\frac{\partial V_1}{\partial z}\right)_{z=0} = \left[\frac{I\rho_1 z}{2\pi(r^2 + z^2)^{\frac{3}{2}}} + \int_0^\infty (-A_1 e^{-mz} + B_1 e^{mz})J_0(mr)mdm\right]_{z=0}$$

$$= \int_0^\infty (B_1 - A_1)J_0(mr)mdm = 0.$$

Since the relation must be true for any value of r, $B_1 - A_1 = 0$, i.e., $B_1 = A_1$, then:

$$V_1 = \frac{I\rho_1}{2\pi}\frac{1}{(r^2 + z^2)^{\frac{1}{2}}} + \int_0^\infty A_1(m)(e^{mz} + e^{-mz})J_0(mr)dm$$

Again, in the last layer, the potential must reduce to zero at $z = \infty$; i.e., $B_n = 0$, and:

$$V_n = \frac{I\rho_1}{2\pi}\frac{1}{(r^2 + z^2)^{\frac{1}{2}}} + \int_0^\infty A_n(m)e^{-mz}J_0(mr)dm$$

The boundary conditions to be satisfied at any boundary are:

$$\begin{aligned} V_i &= V_{i+1} \\ \frac{1}{\rho_i}\frac{\partial V_i}{\partial z} &= \frac{1}{\rho_{i+1}}\frac{\partial V_{i+1}}{\partial z} \quad \text{at } z = H_i \end{aligned} \quad\quad (2.51)$$

We have $2n$ equations, and we have to determine $2n$ unknowns. Thus, the problem can be solved uniquely.
Now:

$$\frac{1}{R} = \frac{1}{(r^2 + z^2)^{\frac{1}{2}}}$$

may be expressed in terms of the well-known Weber's integral formula:

$$\frac{1}{(r^2 + z^2)^{\frac{1}{2}}} = \int_0^\infty e^{-m|z|} J_0(mr)dm \qquad (2.52)$$

and putting $\dfrac{I\rho_1}{2\pi} = q$, we get:

$$V_1 = q\int_0^\infty e^{-m|z|} J_0(mr)dm + \int_0^\infty A_1(m)(e^{mz} + e^{-mz})J_0(mr)dm$$

$$V_i = q\int_0^\infty e^{-m|z|} J_0(mr)dm + \int_0^\infty (A_i(m)e^{-mz} + B_i(m)e^{mz})J_0(mr)dm \qquad (2.53)$$

$$V_n = q\int_0^\infty e^{-m|z|} J_0(mr)dm + \int_0^\infty A_n(m)e^{-mz} J_0(mr)dm$$

From relations (2.51) and (2.53) we get the following systems of equations:

$$\int_0^\infty A_1(e^{mH_1} + e^{-mH_1})J_0(mr)dm = \int_0^\infty (A_2 e^{-mH_1} + B_2 e^{mH_1})J_0(mr)dm$$

$$-\frac{q}{\rho_1}\int_0^\infty e^{-mH_1} J_0(mr)dm + \frac{1}{\rho_1}\int_0^\infty A_1(e^{mH_1} - e^{-mH_1})J_0(mr)dm$$

$$= -\frac{q}{\rho_2}\int_0^\infty e^{-mH_1} J_0(mr)dm + \frac{1}{\rho_2}\int_0^\infty (-A_2 e^{-mH_1} + B_2 e^{mH_1})J_0(mr)dm$$

$$(2.54a)$$

$$\int_0^\infty (A_i e^{-mH_i} + B_i e^{mH_i})J_0(mr)dm = \int_0^\infty (A_{i+1} e^{-mH_i} + B_{i+1} e^{mH_i})J_0(mr)dm$$

$$-\frac{q}{\rho_i}\int_0^\infty e^{-mH_i} J_0(mr)dm + \frac{1}{\rho_i}\int_0^\infty (-A_i e^{-mH_i} + B_i e^{mH_i})J_0(mr)dm$$

$$= -\frac{q}{\rho_{i+1}}\int_0^\infty e^{-mH_i} J_0(mr)dm + \frac{1}{\rho_{i+1}}\int_0^\infty (-A_{i+1} e^{-mH_i} + B_{i+1} e^{mH_i})J_0(mr)dm \ .$$

$$(2.54b)$$

$$\int_0^\infty (A_{n-1} e^{-mH_{n-1}} + B_{n-1} e^{mH_{n-1}}) J_0 (mr) dm = \int_0^\infty A_n e^{-mH_{n-1}} J_0 (mr) dm$$

$$-\frac{1}{\rho_{n-1}} \int_0^\infty (A_{n-1} e^{-mH_{n-1}} + B_{n-1} e^{mH_{n-1}}) J_0 (mr) dm - \frac{q}{\rho_{n-1}}$$

$$\int_0^\infty e^{-mH_{n-1}} J_0 (mr) dm$$

$$= -\frac{q}{\rho_n} \int_0^\infty e^{-mH_{n-1}} J_0 (mr) dm - \frac{1}{\rho_n} \int_0^\infty A_n e^{-mH_{n-1}} J_0 (mr) dm \qquad (2.54c)$$

As the relations (2.54a, b, c) should be valid for all values of r:

$$A_1 (e^{-mH_1} + e^{mH_1}) - A_2 e^{-mH_1} - B_2 e^{mH_1} = 0$$

$$A_1 \rho_2 (e^{mH_1} - e^{-mH_1}) + A_2 \rho_1 e^{-mH_1} - B_2 \rho_1 e^{mH_1} -$$
$$-q(\rho_2 - \rho_1) e^{-mH_1} = 0 \qquad (2.55a)$$

$$A_i e^{-mH_i} + B_i e^{mH_i} - A_{i+1} e^{-mH_i} - B_{i+1} e^{mH_i} = 0$$

$$\rho_{i+1} (-A_i e^{-mH_i} + B_i e^{mH_i}) + \rho_i A_{i+1} e^{-mH_i} - \rho_i B_{i+1} e^{mH_i} -$$
$$-q(\rho_{i+1} - \rho_i) e^{-mH_i} = 0 \qquad (2.55b)$$

$$A_{n-1} e^{-mH_{n-1}} + B_{n-1} e^{mH_{n-1}} - A_n e^{-mH_{n-1}} = 0$$

$$- A_{n-1} \rho_n e^{-mH_{n-1}} + B_{n-1} \rho_n e^{mH_{n-1}} + A_n \rho_{n-1} e^{-mH_{n-1}} -$$
$$-q(\rho_n - \rho_{n-1}) e^{-mH_{n-1}} = 0 \qquad (2.55c)$$

Thus, from the system of eqs. (2.55 a, b, c), it is theoretically possible to find the potential, and hence the field, in any medium. In geoelectric sounding, we are interested in finding only the potential on the surface, and it is sufficient to find out only the coefficient A_1 for some special cases of practical interest.

2.4.1 Homogeneous Earth

$$A_1 = 0$$

and:

$$V = V_0 = \frac{I\rho}{2\pi} \frac{1}{(r^2 + z^2)^{\frac{1}{2}}} = \frac{I\rho}{2\pi} \frac{1}{R}$$

2.4.2 Two-layer Earth

Put $h_2 = \infty$ in Fig. 2.6 then the system of eqs. (2.55 a, b, c) reduce to:

$$A_1 (e^{-mh_1} + e^{mh_1}) - A_2 e^{-mh_1} = 0$$

$$A_1 \rho_2 (-e^{-mh_1} + e^{mh_1}) + A_2 \rho_1 e^{-mh_1} - q(\rho_2 - \rho_1)e^{-mh_1} = 0$$

A simultaneous solution (using Cramer's rule) of the equations give $A_1 (m) = N_1/D$,

where

$$N_1 = \begin{vmatrix} 0 & -e^{-mh_1} \\ q(\rho_2 - \rho_1)e^{-mh_1} & \rho_1 e^{-mh_1} \end{vmatrix} = q(\rho_2 - \rho_1)e^{-2mh_1}$$

$$D = \begin{vmatrix} e^{mh_1} + e^{-mh_1} & -e^{-mh_1} \\ \rho_2(e^{mh_1} - e^{-mh_1}) & \rho_1 e^{-mh_1} \end{vmatrix} = (\rho_2 + \rho_1) + (\rho_2 - \rho_1)e^{-2mh_1}$$

Writing $(\rho_2 - \rho_1)/(\rho_2 + \rho_1) = K_{12}$ we get:

$$A_1(m) = q \frac{K_{12}e^{-2mh_1}}{1 - K_{12}e^{-2mh_1}}$$

$$= q(K_{12}e^{-2mh_1} + K_{12}^2 e^{-4mh_1} + \ldots\ldots + K_{12}^n e^{-2mnh_1} + \ldots\ldots)$$

$$= q \sum_{n=1}^{\infty} K_{12}^n e^{-2mnh_1}$$

Hence:

$$V_1 = q \int_0^{\infty} e^{-mz} J_0(mr)dm + q \sum_{n=1}^{\infty} K_{12}^n \int_0^{\infty} e^{-2mnh_1}(e^{mz} + e^{-mz})J_0(mr)dm$$

Again using Weber's formula (2.52):

$$V_1 = \frac{q}{(r^2 + z^2)^{\frac{1}{2}}} + q \sum_{n=1}^{\infty} K_{12}^n \frac{1}{[r^2 + (2nh_1 - z)^2]^{\frac{1}{2}}} +$$

$$+ q \sum_{n=1}^{\infty} K_{12}^n \frac{1}{[r^2 + (2nh_1 + z)^2]^{\frac{1}{2}}}$$

$$(2.56)$$

Eq. (2.56) gives the potential at any point (r, z) in the first medium. To find the potential on the surface, we put $z = 0$, then:

$$V = \frac{I\rho_1}{2\pi} \left[\frac{1}{r} + 2 \sum_{n=1}^{\infty} \frac{K_{12}^n}{[r^2 + (2nh_1)^2]^{\frac{1}{2}}} \right] \tag{2.57}$$

and the intensity on the surface, $E = -(\partial v/\partial r)$, is:

$$E = \frac{I\rho_1}{2\pi} \left[\frac{1}{r^2} + 2 \sum_{n=1}^{\infty} \frac{K_{12}^n}{[r^2 + (2nh_1)^2]^{\frac{3}{2}}} \right] \tag{2.58}$$

The formulas (2.57) and (2.58) can be used to determine the apparent resistivity for any of the electrode arrangements given in the section on resistivity measurement (2.2).

Considering the Schlumberger arrangement of our interest (for very small MN, i.e., MN→0):

$$\bar{\rho} = 2\pi r^2 \frac{E}{I} = \rho_1 \left[1 + 2 \sum_{n=1}^{\infty} \frac{\delta^3 K_{12}^n}{(\delta^2 + 4n^2)^{\frac{3}{2}}} \right] \tag{2.59}$$

where:

$$\delta = r/h_1 = AB/2h_1$$

2.4.3 Three-layer earth

Put $h_3 = \infty$ in Fig. 2.6 then the set of equations (2.55a, b, c) reduce to:

$$A_1(e^{-mH_1} + e^{mH_1}) - A_2 e^{-mH_1} - B_2 e^{mH_1} = 0$$

$$A_1\rho_2(-e^{-mH_1} + e^{mH_1}) + A_2\rho_1 e^{-mH_1} - B_2\rho_1 e^{mH_1} - $$
$$- q(\rho_2 - \rho_1)e^{-mH_1} = 0$$

$$A_2 e^{-mH_2} + B_2 e^{mH_2} - A_3 e^{-mH_2} = 0 \tag{2.60}$$

$$-A_2\rho_3 e^{-mH_2} + B_2\rho_3 e^{mH_2} + A_3\rho_2 e^{-mH_2} - q(\rho_3 - \rho_2)e^{-mH_2} = 0$$

Solving these equations we get:

$$A_1(m) = q \frac{K_{12} e^{-2mH_1} + K_{23} e^{-2mH_2}}{1 - K_{12} e^{-2mH_1} - K_{23} e^{-2mH_2} + K_{12} K_{23} e^{-2m(H_2 - H_1)}} \tag{2.61}$$

where $K_{23} = (\rho_3 - \rho_2)/(\rho_3 + \rho_2)$

Therefore:

$$V_1 = \frac{I\rho_1}{2\pi}\left[\frac{1}{(r^2+z^2)^{\frac{1}{2}}} + \int_0^\infty \frac{(K_{12}e^{-2mH_1}+K_{23}e^{-2mH_2})(e^{-mz}+e^{mz})}{1-K_{12}e^{-2mH_1}-K_{23}e^{-2mH_2}+K_{12}K_{23}}\right.$$

$$\left.\frac{J_0(mr)dm}{e^{-2m(H_2-H_1)}}\right] \tag{2.62}$$

To express the potential in a convenient form for computation, we proceed in the following way:

Put, $H_1 = p_1H_0$ and $H_2 = p_2H_0$ where p_1 and p_2 are whole numbers, and H_0 has some fixed value.

Then, writing $e^{-2mH_0} = g$, eq. (2.61) may be written as:

$$A_1(m) = q\frac{K_{12}g^{p_1}+K_{23}g^{p_2}}{1-K_{12}g^{p_1}-K_{23}g^{p_2}+K_{12}K_{23}g^{(p_2-p_1)}} \tag{2.63}$$

As p_1 and p_2 are whole numbers, $A_1(m)$ is a rational function of g, that is:

$$A_1(m) = q(b_1g+b_2g^2+b_3g^3+\ldots\ldots) = q\sum_{n=1}^\infty b_ng^n = q\sum_{n=1}^\infty b_ne^{-2mnH_0} \tag{2.64}$$

Comparing eqs. (2.63) and (2.64):

$$K_{12}g^{p_1}+K_{23}g^{p_2} = \left[1-K_{12}g^{p_1}-K_{23}g^{p_2}+K_{12}K_{23}g^{(p_2-p_1)}\right]\sum_{n=1}^\infty b_ng^n \tag{2.65}$$

This identity requires that the coefficients of any order of g must be identically equal on both sides. As the highest order on the left hand side is g^{p_2}, the coefficient of g of an order greater than p_2 on the right-hand side must be zero. Let us write down the coefficients of g^{p_2+t}, where t is a positive number, then:

$$b_{p_2+t} - K_{12}b_{p_2-p_1+t} - K_{23}b_t + K_{12}K_{23}b_{p_1+t} = 0$$

gives the recurrence formula:

$$b_{p_2+t} = K_{12}b_{p_2-p_1+t} + K_{23}b_t - K_{12}K_{23}b_{p_1+t} \tag{2.66}$$

Thus, b_{p_2+t} may be calculated, knowing the values of $b_{p_2-p_1+t}$, b_t and b_{p_1+t}. The coefficients up to the maximum value b_{p_2}, may be determined from eq. (2.65). The rest of the coefficients can be determined through the use of the recurrence formula (2.66).

Thus, the potential at any point in the first layer can be written as:

$$V_1 = \frac{I\rho_1}{2\pi}\left[\frac{1}{(r^2 + z^2)^{\frac{1}{2}}} + \sum_{n=1}^{\infty}\frac{b_n}{[r^2 + (2nH_0 + z)^2]^{\frac{1}{2}}} + \right.$$

$$\left. + \sum_{n=1}^{\infty}\frac{b_n}{[r^2 + (2nH_0 - z)^2]^{\frac{1}{2}}}\right]$$

At the surface z = 0:

$$V = \frac{I\rho_1}{2\pi}\left[\frac{1}{r} + 2\sum_{n=1}^{\infty}\frac{b_n}{[r^2 + (2nH_0)^2]^{\frac{1}{2}}}\right] \tag{2.67}$$

and:

$$E = \frac{I\rho_1}{2\pi}\left[\frac{1}{r^2} + 2\sum_{n=1}^{\infty}\frac{b_n r}{[r^2 + (2nH_0)^2]^{\frac{3}{2}}}\right] \tag{2.68}$$

Thus, for the Schlumberger symmetrical arrangement:

$$\bar{\rho} = \rho_1\left[1 + 2\sum_{n=1}^{\infty}\frac{b_n r^3}{[r^2 + (2nH_0)^2]^{\frac{3}{2}}}\right]$$

In practice, the thickness of the second layer is usually expressed in terms of the thickness of the first layer, i.e., $h_1 (= H_1)$. Then ;

$$\bar{\rho} = \rho_1\left[1 + 2\sum_{n=1}^{\infty}\frac{b_n r^3}{[r^2 + (2nH_1)^2]^{\frac{3}{2}}}\right]$$

or if $r/h_1 = \delta$:

$$\bar{\rho} = \rho_1\left[1 + 2\sum_{n=1}^{\infty}\frac{b_n \delta^3}{(\delta^2 + 4n^2)^{\frac{3}{2}}}\right] \tag{2.69}$$

2.4.4 Four-layer Earth

In the case of a four-layer earth, it can be shown that:

$$A_1(m) = \frac{q(K_{12}g^{p_1} + K_{23}g^{p_2} + K_{34}g^{p_3} + K_{12}K_{23}K_{34}g^{p_3-p_2-p_1})}{1 - K_{12}g^{p_1} - K_{23}g^{p_2} - K_{34}g^{p_3} + K_{12}K_{23}g^{p_2-p_1} +}$$

$$+ K_{23}K_{34}g^{p_3-p_2} + K_{12}K_{34}g^{p_3-p_1} - K_{12}K_{23}K_{34}g^{p_3-p_2+p_1}$$

$$(2.70)$$

Here, also, the potential can be expressed in the same form as eq.(2.67) and the apparent resistivity expressed in the form eq.(2.69). A recurrence relation for $A_1(m)$, known as Kernel function, may be evolved for an n-layer earth in line with Flathe (1955, p.272, eq.3) but this is not likely to be useful in view of the development of rapid computational approaches detailed later.

The complications in computation increase with the increase in number of layers. With the use of computers it has been possible to plot sets of theoretical master curves to be used for interpretation and such sets of two, three- and four-layer master curves are available in published form (Mooney and Wetzel, 1956; Compagnie Generale de Geophysique, 1963; Orellana and Mooney, 1966).

It may be mentioned here that various simplified approaches to the computation of theoretical curves have been suggested from time to time. Flathe(1955) introduced his method of calculating sounding curves with an ordinary desk calculator; but this is suitable only for cases approximated by a perfectly conducting and perfectly insulating substratum. Van Dam (1965) has introduced a simple method for the calculation of sufficiently exact sounding curves with hand calculators.

A procedure to compute apparent resistivity curves for layered earth structure for the Schlumberger, Wenner and dipole configurations, has been given by Mooney et al. (1966), where use is made of large digital computers. In this method, the formulation is relatively simple, and a program can handle any number of layers; in addition, a single set of stored coefficients can be used repeatedly by different electrode spacing and for different electrode arrangements. The technique is claimed to be relatively simpler and more accurate compared to those described by the Compagnie Generale de Geophysique (CGG 1955, 1963) and Flathe (1955). However, the method suggested by Van Dam (1967) is somewhat similar to the one described by Mooney et al. (1966), and this method is meant for use with digital calculators.

Existing methods (CGG, 1963; Flathe, 1955; Mooney et al., 1966) based on evaluation of Kernel function requires that layer thicknesses be multiples of some common thickness for a rapid convergence of the series to be computed. Calculation of apparent resistivity curves with known layer parameters using inverse filter coefficients (Ghosh, 1971a), on the other hand, is without such restriction and is straightforward. The method will be explained in Chapter 4.

2.5 Principle of Equivalence

In this section we shall give the theoretical basis of a result which, as we shall see later, serves a very important purpose in the interpretation of geoelectric sounding curves. This is known as the "principle of equivalence".

We know that the potential at the surface of a three-layer earth depends on the coefficient A_1 given by:

$$A_1(m) = q \frac{K_{12}e^{-2mh_1} + K_{23}e^{-2m(h_1+h_2)}}{1 - K_{12}e^{-2mh_1} - K_{23}e^{-2m(h_1+h_2)} + K_{12}K_{23}e^{-2mh_2}} \quad (2.71)$$

where $q = \dfrac{I\rho_1}{2\pi}$; K_{12} and K_{23} have usual meaning.

Case I

Let us suppose $h_2 \ll h_1$; $\rho_2 \ll \rho_1$ and $\rho_3 \gg \rho_2$, then:

$$e^{-2m(h_1+h_2)} = e^{-2mh_1} + e^{-2mh_2} \approx e^{-2mh_1}(1 - 2mh_2)$$

$$K_{12} = [2\rho_2/(\rho_2+\rho_1) - 1] \sim (2\rho_2/\rho_1) - 1$$

$$K_{23} = 1 - [2\rho_2/(\rho_3+\rho_2)] \approx 1 - (2\rho_2/\rho_3)$$

Eq. (2.71) can be written as:

$$A_1(m) = qe^{-2mh_1} \frac{\rho_3 - \rho_1 - m\rho_1\rho_3(h_2/\rho_2)}{-\left[\left(\dfrac{1}{\rho_1} - \dfrac{1}{\rho_3}\right) - m(h_2/\rho_2)\right]e^{-2mh_1} + \dfrac{1}{\rho_1} - \dfrac{1}{\rho_3} + m(h_2/\rho_2)}$$

$$(2.72)$$

$$= qe^{-2mh_1} \frac{\rho_3 - \rho_1 - m\rho_1\rho_3 S}{-(\rho_3 - \rho_1 - m\rho_1\rho_3 S)e^{-2mh_1} + \rho_3 - \rho_1 + m\rho_1\rho_3 S}$$

It is seen from the relation (2.72) that $A_1(m)$ does not depend on the absolute values of ρ_2 and h_2, but only on the ratio $h_2/\rho_2 = S$.

Case II

Similarly, when $h_2 \ll h_1$; $\rho_2 \gg \rho_3$ and $\rho_2 \gg \rho_1$:

$$K_{12} = 1 - \frac{2\rho_1}{\rho_2 + \rho_1} \approx 1 - \frac{2\rho_1}{\rho_2}$$

$$K_{23} = \frac{2\rho_3}{\rho_2 + \rho_3} - 1 \approx \frac{2\rho_3}{\rho_2} - 1$$

and:

$$A_1(m) = qe^{-2mh_1} \frac{\rho_3 - \rho_1 + mT}{(\rho_1 - \rho_3 - mT)e^{-2mh_1} + \rho_1 + \rho_3 + mT} \qquad (2.73)$$

Here A_1(m) depends only on $T = h_2\rho_2$ and not individually on ρ_2 and h_2.

Case I refers to an H-type section and case II refers to a K-type section, as will be explained later.

According to the above discussion, it can be said that H-type curves are "equivalent" with respect to S, provided the intermediate layer has a thickness and resistivity which is very small compared to those of the other two; and K-type curves are "equivalent" with respect to T if the thickness of the intermediate layer is small, but resistivity is large, compared to the other two layers.

The "principle of equivalence" imposes restriction in the interpretation of sounding data by introducing error, in precise determination of thickness, small compared to depth. This "ambiguity" is removed through geological and other controls.

Similarly, we face difficulty in detecting beds having resistivity values intermediate to those of the enclosing beds. If the bed is not considerably thick its effect will not be reflected on the apparent resistivity curve. When the thickness of the bed increases it has to compensate the effect due to the increase in resistivity of the enclosing beds, controlled by the "principle of suppression".

2.6 Vertical Electrical Sounding (VES)

It is seen that the apparent resistivity as measured on the surface of an inhomogeneous earth is given by:

$$\bar{\rho} = K(\Delta V/I)$$

Where K, a geometrical factor, depends on the configuration of the current as well as measuring electrodes. Broadly speaking, we can distinguish two types of resistivity measurements. In the first, known as "geoelectric profiling" or mapping, the value of K remains constant for a particular set of reading, and measurements are carried out by laterally shifting the electrodes (current and potential) along with the centre of the electrode array. In this way we get the variation in resistivity along a predetermined line and at a fixed depth.

In the second method known as "geoelectric sounding", for a particular set, the distances between the current electrodes are changed (increased gradually) keeping the centre of the spread fixed. Here, the value of K progressively changes. In this way, the apparent resistivity values at the surface reflect the vertical distribution of resistivity values in a geological section. This is why geoelectric sounding is sometimes known as "vertical electrical drilling". The present book deals with geoelectric sounding using only Schlumberger electrode arrangement referred to as "vertical electrical sounding" and abbreviated, here as VES. The present book deals with VES using symmetrical electrodes arrangement (Fig. 2.2).

An arrangement of electrodes which is widely used is the symmetrical one comprising Schlumberger and Wenner arrays. Here, the current electrodes A, B are symmetrically placed with respect to the potential electrodes M, N (Fig. 2.2), and where the centre of the spread, O, is the sounding point. In the Schlumberger method, dealt with in this book, the sounding may be done (theoretically speaking) by moving only the current electrodes, progressively increasing the distance AB. However, when AB is large compared to MN, the potential drop between M and N may be too small to be measured.

Hence, in practice it is necessary also to increase the distance between M and N, whenever required, depending on the sensitivity of the measuring instrument. The value of MN and the corresponding values of AB are chosen in order to get overlapping readings whenever a change-over of MN from one value to the other takes place. The usual distribution in the values of MN and AB for an average instrument sensitivity and for a maximum spread $AB = 1000$ m, is tabulated in Table 2.1. This is subject to modification, depending on the field conditions, instrument sensitivity, and other practical difficulties. In the beginning, a few readings may be taken with the Wenner arrangement, so as to get the resistivity of the surface layer.

In any of these observations for corresponding values of $MN/2$(m) and $AB/2$(m), resistance (ohms) is calculated from $\Delta V(mV)$ and I (mA). The resistance is multiplied by the corresponding value of K (Table 2.1) to get the apparent resistivity value in ohm-m.

Table 2.1 A Typical field layout for Schlumberger arrangement

Obs. No.	MN/2 (m)	AB/2 (m)'	K	Obs. No.	MN/2 (m)	AB/2 (m)	K
1	0.5	1.5	6.28	18	..	60	549.5
2	..	2	11.8	19	..	80	989.1
3	..	3	27.5	20	..	100	1554.3
4	..	4	49.4	21	20	100	753.6
5	1	4	23.5	22	..	120	1099
6	..	6	54.9	23	..	140	1502
7	..	8	99.0	24	..	160	1978
8	..	10	155.0	25	..	180	2512
9	2	10	75.0	26	..	200	3100
10	..	15	173.0	27	40	200	1507
11	..	20	310.0	28	..	250	2402
12	5	20	118.5	29	..	300	3470
13	..	25	188.5	30	..	350	4824
14	..	30	274.8	31	..	400	6217
15	..	40	494.5	32	..	500	9750
16	..	50	777.0	33	80	400	3485
17	10	50	376.8	34	..	500	4781

2.7 Approximate Computational Procedure

It has been shown in eq. (2.69) that for the Schlumberger arrangement:

$$\bar{\rho} = \rho_1 \left[1 + 2 \sum_{n=1}^{\infty} \frac{b_n \delta^3}{\{\delta^2 + (2n)^2\}^{\frac{3}{2}}} \right]$$

where the coefficient b_n is given by eq. (2.65); for $\rho_2 > \rho_1$ this is given by the recurrence formula represented by eq. (2.66). Once b_n is known, the value of $\bar{\rho}/\rho_1$ can easily be calculated for various values of δ (i.e., r/h_1) with the help of a computer, when expressed in the following way:

$$\bar{\rho}/\rho_1 = 1 + \frac{2b_1 \delta^3}{(\delta^2 + 4)^{\frac{3}{2}}} + \frac{2b_2 \delta^3}{(\delta^2 + 16)^{\frac{3}{2}}} + \ldots + \ldots$$

In the case of a three-layer problem, the computations are simplified when $\rho_3 = 0$ or ∞, because in that case the coefficient $A_1(m)$ given in eq. (2.63) is further simplified.

Let us consider a specific case with the following values:

$$\rho_1 = 100 \text{ ohm-}m; \quad \rho_2 = 5 \text{ ohm-}m; \quad \rho_3 = \infty$$

$$h_1 = 10 \ m; \qquad h_2 = 10 \ m; \qquad h_3 = \infty$$

Then $H_1 = h_1$ and $H_2 = 2h_1$, i.e.,

$$p_1 = 1 \text{ and } p_2 = 2.$$

From eq. (2.63) comparing with eq. (2.64), we can write the value of the coefficient $A_1(m)$ given by:

$$A_1 = \frac{K_{12} g + g^2}{1 - g^2} = \sum_{n=1}^{\infty} b_n g^n \qquad (2.74)$$

or:

$$A_1 = a \frac{g}{1-g} + c \frac{-g}{1-g}$$

such that:

$$(a - c) = K_{12};$$
$$(a + c) = 1,$$

or:

$$A_1 = a \frac{g}{1-g} + c \frac{-g}{1-g}$$

On expansion into a series:

$$A_1 = a(g + g^2 + g^3 + \ ......) + c(-g - g^2 - g^3 - \ ......) = \sum_{n=1}^{\infty} b_n g^n \qquad (2.75)$$

This gives $b_1 = b_2 = b_3 = 1$ for the first part, and $b_1 = b_2 = b_3 = (-1)$ for the second part, and the apparent resistivity $\bar{\rho}$ is given by:

$$\bar{\rho}/\rho_1 = 1 + \frac{1 + K_{12}}{2} \cdot 2 \sum_{n=1}^{\infty} \frac{\delta^3}{\{\delta^2 + (2n)^2\}^{\frac{3}{2}}} + \frac{1 - K_{12}}{2} \cdot 2 \sum_{n=1}^{\infty} \frac{\delta^3 (-1)^n}{\{\delta^2 + (2n)^2\}^{\frac{3}{2}}}$$

$$(2.76)$$

The value of $\bar{\rho}/\rho_1$ can be computed from the above expression for a certain value of K_{12}, i.e., (ρ_2/ρ_1) for different value of $\delta = r/h_1 = AB/2h_1$.

In the present case, $K_{12} = 0.905$ and the variables are δ and n. The value of n should be chosen properly, so as to increase the possibility of neglecting the effect of higher order terms through the computer program without appreciable error in the calculation. For a highly convergent series, sufficient accuracy may be obtained within a few tens of terms.

However, for some unfavourable conditions it may be necessary to consider large number terms to be taken care of by the software.

The above relation (2.76), giving the value of $\bar{\rho}/\rho_1$, presents a very interesting point; and it can be further simplified, giving rise to the possibility of using two-layer curves for plotting the three-layer curve.

In the case of two-layer earth, we know that for $\rho_2 = \infty$, the value of $\bar{\rho}$ is given by:

$$\bar{\rho}_\infty = \rho_1 \left[1 + 2 \sum_{n=1}^{\infty} \frac{\delta^3}{\{\delta^2 + (2n)^2\}^{\frac{3}{2}}} \right]$$

and, similarly, for $\rho_2 = 0$, we can write:

$$\bar{\rho}_0 = \rho_1 \left[1 + 2 \sum_{n=1}^{\infty} \frac{\delta^3 (-1)^n}{\{\delta^2 + (2n)^2\}^{\frac{3}{2}}} \right]$$

Thus, eq. (2.76) may be written as:

$$\bar{\rho} = \frac{1 - K_{12}}{2} \bar{\rho}_0 + \frac{1 + K_{12}}{2} \bar{\rho}_\infty \tag{2.77}$$

Therefore, eq.(2.77) indicates that we can plot the three-layer theoretical master curves for the given case with the help of the two-layer master curves available for $\rho_2 = 0$ and $\rho_2 = \infty$. This forms the basis of graphical construction of three- and four-layer curves for use in the field where computers may not be available.

3 Interpretation of Data

The interpretation of Schlumberger Vertical Electrical Sounding (VES) data requires a large number of theoretical master curves for comparison with field curves. These field curves are now generated as required with the computers. The well-known partial curve matching technique using available album of master curves for interpretation of layer parameters is always used at least as the initial guess for the direct interpretation methods for inversion of VES data.

This chapter deals with the basic principles and procedures for interpretation of Schlumberger VES curves using both conventional curve matching and latest techniques on inversion of resistivity field data. In the following section, we deal with the "type curves" with VES obtained through field surveys.

3.1 Schlumberger Apparent Resistivity Type Curves

3.1.1 Two-layer Curves

Two sets of theoretical two-layer master curves are available for (ρ_2/ρ_1) greater than unity, ascending type (Set I) and for (ρ_2/ρ_1) less than unity descending type (Set II). The values of ρ_2/ρ_1 for which curves have been plotted are:

Set I : ρ_2/ρ_1 = 11/9, 3/2, 13/7, 2, 7/3, 3, 4, 5, 17/3, 7, 9, 19, 39, 99, ∞

Set II : ρ_2/ρ_1 = 9/11, 2/3, 7/13, 1/2, 3/7, 1/3, 1/4, 1/5, 3/17, 1/7, 1/9, 1/19, 1/39, 1/99, 0.

These sets of theoretical master curves have been reproduced in Figs. 3.1(a and b) plotted on a double-logarithm graph sheet with a modulus of 62.5 mm (Bhattacharya and Patra, 1968) and can be used for construction and interpretation of multi-layer curves.

3.1.2 Three-layer Curves

The whole set of three-layer sounding curves can be divided into four groups, depending on the relative values of ρ_1, ρ_2 and ρ_3:

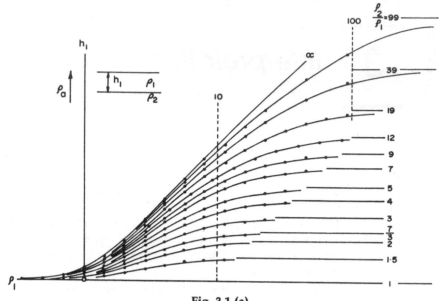

Fig. 3.1 (a)

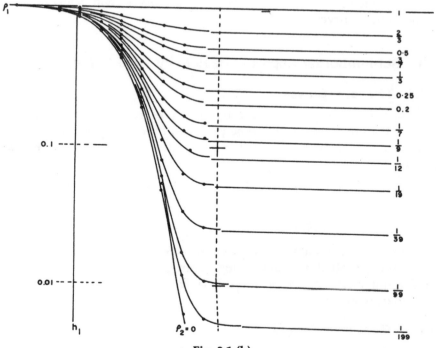

Fig. 3.1 (b)

1. Minimum type: when $\rho_1 > \rho_2 < \rho_3$. This is referred to as H-type (associated with the name of Hummel).
2. Double ascending type : when $\rho_1 < \rho_2 < \rho_3$. This is also known as A-type (corresponding to the term anisotropy).
3. Maximum type : when $\rho_1 < \rho_2 > \rho_3$. This is known as K-type or is sometimes referred to as DA-type (meaning displaced or modified anisotropy).
4. Double descending type : when $\rho_1 > \rho_2 > \rho_3$. This is known as Q-type and is sometimes referred to as DH-type (meaning displaced Hummel or modified Hummel).

A diagrammatical representation of all these type curves is given in Fig. 3.2 for the three-layer cases. The following theoretical three-layer master curves for Schlumberger configuration are available in published form: (1) The Compagnie General de Geophysique (1955, 1963) contains 48 sets of curves, each set containing 10 curves giving a total of 480

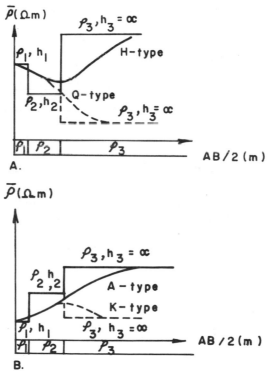

Fig. 3.2 Three-layer type curves.
    A. H - type ($\rho_1 > \rho_2 < \rho_3$) and Q-type ($\rho_1 > \rho_2 > \rho_3$)
    B. A - type ($\rho_1 < \rho_2 < \rho_3$) and K-type ($\rho_1 < \rho_2 > \rho_3$)

separate curves available for interpretation, (2) Orellana and Mooney (1966) present master tables and curves representing 76 three-layer sets (25 each of H and K-type and 13 each of Q and A-type) with a total of 912, (3) EAEG (1969) publication of standard set of three-layer curves has been an attractive addition with 2260 cases (prepared by Rijkswaterstaat). The parameters of the curves are noted on the different sets of the album of curves.

3.1.3 Four-layer Curves

From a combination of the curves of the types H, A, K and Q (Fig. 3.2), it is easily seen that there can be only eight types of four-layer curves, shown in Fig. 3.3. These may be designated as HA, HK; AA, AK; KH, KQ and QH, QQ. Theoretically plotted master curves for four-layer cases are available as "Paletka" in Anonymous (1963). The album of four-layer theoretical master curves contains 122 sets, covering all the eight types.

The values of the parameters are given below:

$\rho_2/\rho_1 = 1/39, 1/19, 1/9, 3/17, 1/4, 3/7, 2/3, 3/2, 7/3, 17/3, 3, 4, 9$ and 39;

$\rho_3/\rho_1 = 1/39, 1/19, 1/9, 3/17, 1/4, 3/7, 2/3, 3/2, 7/3, 17/3, 3, 4, 9$ and 39;

$h_2/h_1 = 1/2, 1, 2, 3, 5, 24$;

$h_3/h_1 = 1/2, 1, 2, 3, 10, 12$, and 72.

The four-layer curves published by Orellana and Mooney (1966) consist of a total of 480 cases distributed in 30 sets.

3.2 Asymptotic Values of Schlumberger Curves

The apparent resistivity ($\bar{\rho}$) for a two-layer earth may be written, for Schlumberger arrangement (Bhattacharya and Patra, 1968, eq. 2.59) as:

$$\frac{\bar{\rho}}{\rho_1} = 1 + 2 \sum_{n=1}^{\infty} \frac{K_{12}^n (AB/2h_1)^3}{[(AB/2h_1)^2 + (2n)^2]^{\frac{3}{2}}} \tag{3.1}$$

where:

$K_{12} = (\rho_2 - \rho_1)/(\rho_2 + \rho_1)$
AB = electrode separation
h_1 = thickness of the first layer.

Several limiting cases which can be derived from eq. (3.1) are:
(a) When $\rho_2 = \rho_1$, $\bar{\rho} = \rho_1$, the apparent resistivity is equal to the true resistivity of the semi-infinite medium.

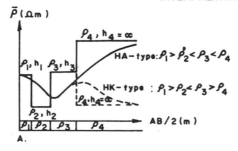

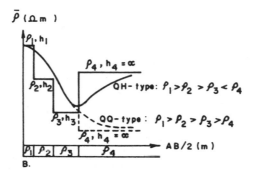

A.

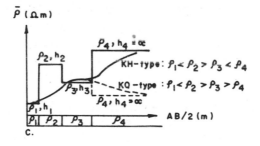

B.

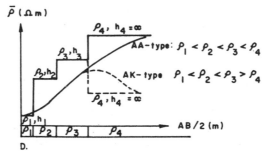

C.

D.

Fig. 3.3 Nature of four-layer type curves. A. HA - and HK - type; B. QH - and QQ - type; C. KH–and KQ–type; D. AA - and AK - type.

(b) When $AB/2 \to 0$, $\bar{\rho} = \rho_1$, i.e., for a two-layer earth the apparent resistivity is equal to the true resistivity of the first layer for small values of electrode separation.

(c) When $AB/2 \to \infty$, $\bar{\rho} = \rho_2$, i.e., the apparent resistivity is equal to the true resistivity of the second layer for large values of electrode separation.

We can write eq.(3.1) in the form:

$$\bar{\rho} = \rho_1 \, F(AB/2h_1) \tag{3.2}$$

assuming that K_{12}, i.e., ρ_2/ρ_1 remains constant.

If we plot $\bar{\rho}$ against $AB/2h_1$ from eq.(3.2) on an arithmetic scale, we get different curves for different values of ρ_1—even for a fixed value of h_1—and similarly for each value of h_1 with ρ_1 fixed.

Using logarithmic scale for eq.(3.2) the influence of ρ_1 and h_1 on the form of the curve may be removed as we get:

$$\log \bar{\rho} - \log \rho_1 = F(\log AB/2 - \log h_1) \tag{3.3}$$

or:

$$\log(\bar{\rho}/\rho_1) = F(\log AB/2h_1) \tag{3.4}$$

Eq.(3.4) shows that a plotting of $\bar{\rho}$ (ordinate) and $AB/2$ (abscissa) on a double-logarithmic scale will give curves of exactly the same form for any value of ρ_1 and h_1 as long as $\rho_2/\overset{*}{\rho}_1$ remains constant. The effect, then, of ρ_1 is to shift the curve upward or downward parallel to the ordinate, and that of h_1 is to shift it to the left or right, parallel to the abscissa (Fig. 3.4).

Thus, the form of the Schlumberger curves, plotted on a double-logarithmic scale, is independent of the resistivity and thickness of the first layer in a two-layer section if ρ_2/ρ_1 is constant. This is found to be valid for a multi-layer geoelectric section also.

From field measurements we get apparent resistivity as a function of the electrode separation, i.e., $\bar{\rho} = f(AB/2)$.

Using the logarithm scale for this relation we get:

$$\log \bar{\rho} = F(\log AB/2) \tag{3.5}$$

Eqs.(3.3) and (3.5) are of the form : $y - b = f(x - a)$, and : $y = f(x)$.

These equations are similar to each other, except that the first curve is shifted parallel to the coordinates with respect to the second curve plotted on the logarithm scale. This shows that the interpretation of the field curves by matching is made possible through the use of the logarithmic scale. Thus, for each value of ρ_1 and h_1, we need not have different curves as form of the curve remains unchanged in log scale, a single

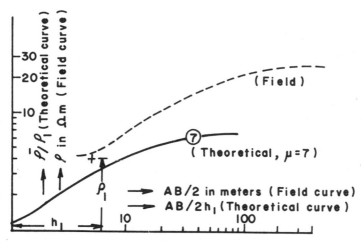

Fig. 3.4 Relation between theoretical and field curves. Field curve of the same form as the theoretical one but shifted with respect to it and parallel to the Co-ordinate axes.

master curve may be used for any value of ρ_1 and h_1, provided ρ_2/ρ_1 remains the same.

In Fig. 3.5, the two layer field curve ($\bar{\rho}$ vs. $AB/2$) is shown for $h_1= 8$ m and $\rho_1 = 4$ ohm-m and $\rho_2/\rho_1 = 7$. The two-layer theoretical master curve ($\bar{\rho}/\rho_1$ vs. $AB/2h_1$) for $\rho_2/\rho_1 = 7$ (the same as the field curve) is shown in the same diagram. These two curves can be matched easily by shifting the field curve over the theoretical curve, keeping the axes parallel. Actually, the theoretical curve can be matched with any field curve for any value of ρ_1 and h_1, provided ρ_2/ρ_1 is the same as that of the theoretical curve.

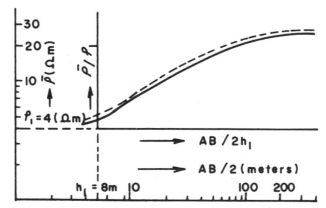

Fig. 3.5 A two-layer field curve superimposed over a two - layer master curve. Origin of the master curve as read over the field curve gives the thickness and resistivity of the upper layer.

The method adopted for finding ρ_1 and h_1 has been indicated in Fig. 3.4. The procedure is to plot the field curve on a transparent double-logarithmic graph sheet which has a modulus the same as that for the theoretical master curves (a modulus of 62.5 mm.), and then to superpose the transparent graph sheet on the master curve; the transparent graph sheet is moved parallel to the coordinates until a match is obtained. Figure 3.5 represents the matched condition, and the point on the transparent double-logarithmic sheet coinciding with the origin of the master curve ($\bar{\rho}/\rho_1 = 1$, $AB/2h_1 = 1$) gives, along the abscissa, $\log(AB/2)$ = $\log h_1$, i.e., $h_1 = AB/2$ (m), and along the ordinate it gives $\log\bar{\rho} = \log\rho_1$ i.e., $\rho_1 = \bar{\rho}$ (ohm-m).

Thus, the use of the logarithm scale opens up the possibility of determining ρ_1 and h_1 from the theoretical and field curves.

Besides, as large variation in resistivity and larger spacings are to be accommodated, log scale is the only logical choice. Further, log scale suppresses effect of high resistivity and thin layers at larger depths but enhances low resistivity and thin layers at smaller depths which is, obviously, advantageous.

It is easily seen that on a double-logarithm scale, the conditions of limiting values are still satisfied and the asymptotic nature is retained, and therefore, double-logarithmic scale is most useful.

Let us now find the asymptotic values of the apparent resistivity when the second layer is of infinite resistivity (basement). It is obvious that at sufficiently large distances from the sources, the current lines will be all parallel to the surface and the equipotential surfaces will be cylindrical, having the vertical through the sources as the axis. Let us consider an equipotential surface at a large distance r; then the current, I, is given by $I = 2\pi r h_1 J$, where $J = E/\rho_1$.

Therefore:

$$E = \rho_1 I/2\pi r h_1$$

Hence, the apparent resistivity for Schlumberger configuration is given by:

$$\bar{\rho} = 2\pi r^2 \, (E/I) = (\rho_1/h_1)r$$

Now taking the logarithm, we get:

$$\log\bar{\rho} = \log r + \log(\rho_1/h_1) = \log r - \log (h_1/\rho_1) \tag{3.6}$$

This is the equation of a straight line inclined at an angle of 45° to the abscissa, cutting it at a distance (h_1/ρ_1) from the origin. It can be shown that for an n-layer earth (n^{th} layer of infinite resistivity) the same asymptotic relation holds good provided ρ_1 is changed to ρ_s - longitudinal resistivity of $(n-1)$ layers and h_1 is changed to H - the total thickness of $(n-1)$ layers.

3.3 Principle of Reduction

Consider a prism of unit cross-section, with thickness h and resistivity ρ. Then the resistance (T) normal to the face of the prism, and the conductance (S) parallel to the face of the prism, are given by:

$$T = h\rho \tag{3.7}$$

and:

$$S = h/\rho \tag{3.8}$$

wherefrom we get:

$$h = \sqrt{ST} \quad \text{and} \quad \rho = \sqrt{T/S} \tag{3.9}$$

Thus, each value of S and T determines a section with definite values of h and ρ given by eq.(3.9). Now, from eq.(3.7) we can write:

$$\log \rho = -\log h + \log T \tag{3.10}$$

This equation defines a straight line inclined at an angle of 135° to the h-axis and cutting it at a distance T from the origin, if ρ is plotted against h on a double-logarithm scale.

Similarly, from relation (3.8) we get:

$$\log \rho = \log h - \log S \tag{3.11}$$

which defines a straight line—also see eq. (3.6)—inclined at an angle of 45° with abscissa (h-axis) and meeting it at a distance S from the origin.

The point of intersection of the two straight lines defined by eqs. (3.10) and (3.11) then uniquely defines the resistivity and thickness for a particular combination of T and S.

Consider now that the prism consists of n parallel homogeneous and isotropic layers of resistivities $\rho_1, \rho_2, ..., \rho_n$ and thicknesses $h_1, h_2, ..., h_n$ respectively (Fig. 3.6).

When the current is flowing normal to the base, the total resistance of the prism is:

$$T = T_1 + T_2 + ... + T_n = \sum_{i=1}^{n} T_i = \rho_1 h_1 + \rho_2 h_2 + ... + \rho_n h_n = \sum_{i=1}^{n} \rho_i h_i \tag{3.12}$$

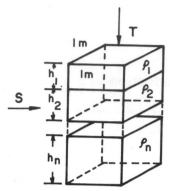

Fig. 3.6 An n-layer prism of unit cross-section.

$$S = S_1 + S_2 + ... + S_n = \sum_{i=1}^{n} S_i = h_1/\rho_1 + h_2/\rho_2 + ... + h_n/\rho_n = \sum_{i=1}^{n} h_i/\rho_i \quad (3.13)$$

The parameters T and S defined as transverse resistance and longitudinal conductance, respectively, play a very important role in the interpretation of sounding data. It may be mentioned that Maillet (1947) used the notations R and C for these parameters, and called them "Dar Zarrouk Variable" and "Dar Zarrouk Function", respectively. Zohdy (1973, 1974) has utilized Dar Zarrouk curves for interpretation of Schlumberger curves through a computer program which automatically calculates layer parameters from apparent resistivity curves.

For the particular case of a two-layer prism, we get:

$$T = T_1 + T_2 = \rho_1 h_1 + \rho_2 h_2 \quad (3.14)$$

and

$$S = S_1 + S_2 = h_1/\rho_1 + h_2/\rho_2 \quad (3.15)$$

If ρ_s and ρ_t are, respectively, the longitudinal and transverse resistivities of the block, then:

$$\rho_t(h_1 + h_2) = \rho_1 h_1 + \rho_2 h_2 \quad (3.16)$$

and

$$(h_1 + h_2)/\rho_s = h_1/\rho_1 + h_2/\rho_2 \quad (3.17)$$

Thus, the coefficient of anisotropy λ and the mean resistivity ρ_m are respectively given by:

$$\lambda = \sqrt{\rho_t/\rho_s} = \frac{1}{h_1 + h_2}[(h_1\rho_1 + h_2\rho_2)(h_1/\rho_1 + h_2/\rho_2)]^{\frac{1}{2}} \quad (3.18)$$

and:

$$\rho_m = \left[\frac{h_1 \rho_1 + h_2 \rho_2}{h_1 / \rho_1 + h_2 / \rho_2} \right]^{\frac{1}{2}} \tag{3.19}$$

We shall now assume that the anisotropic prism may be replaced by a homogeneous and isotropic prism of thickness h_e, and resistivity ρ_e, which may be called, respectively, the effective thickness and the effective resistivity of the block.
Then,

$$h_e \rho_e = T = h_1 \rho_1 + h_2 \rho_2 \tag{3.20}$$

$$h_e \rho_e = S = h_1 / \rho_1 + h_2 / \rho_2 \tag{3.21}$$

from which we get:

$$h_e = [(h_1 \rho_1 + h_2 \rho_2)(h_1 / \rho_1 + h_2 / \rho_2)]^{\frac{1}{2}} = \lambda(h_1 + h_2) = \lambda H \tag{3.22}$$

$$\rho_e = \left[\frac{h_1 \rho_1 + h_2 \rho_2}{h_1 / \rho_1 + h_2 / \rho_2} \right]^{\frac{1}{2}} = \rho_m = \lambda \rho_s \tag{3.23}$$

Thus, it is possible to transform an isolated two-layer block (each homogeneous and isotropic) into a single homogeneous and isotropic medium. A complete isolation of this kind is possible in the case of A-type curves, where the third layer is highly resistive and the second layer is more resistive than the first. Here, the effective thickness of the reduced layer is equal to λ times the total thickness and the effective resistivity is equal to the mean resistivity of the mediums. Since λ is always greater than unity, the effective thickness of the composite layer is greater than the total thickness of the two layers.

3.4 Principle or Reduction of a Three-layer Earth

We shall now consider the case of three-layer earth. The left-hand part of the sounding curve in this case coincides with a two-layer curve having a parameter $\mu = \rho_2 / \rho_1$; the closer the coincidence, the smaller the value of the spread AB. For larger values of AB, the right-hand part of the curve may be taken to coincide with a two-layer curve having a parameter $\mu_e = \rho_3 / \rho_e$ where ρ_3 is the resistivity of the third layer and ρ_e some effective resistivity of a homogeneous layer replacing the first two layers. Our problem now is to find the parameters (the effective resistivity and effective thickness) of this reduced layer. From a careful study of a large number of theoretical, experimental and field curves, it has been possible to obtain some empirical laws for determining these parameters. These laws are different for the four different types of the three-layer curves and have been formed to be useful in the construction and interpretation of field curves.

These cases will now be considered separately.

3.4.1 Case I: H-type

In this case, the resistivity of the intermediate layer is lower than that of the top and the bottom layers. When the resistivity of the bottom layer is large, the sounding curve is appreciably affected by the resistive substratum for large values of AB, and the flow of current in the upper layers will be approximately parallel to the horizontal strata. Thus, in this case, the transverse resistance (T) is negligible; and the longitudinal conductance (S) is the sum of the longitudinal conductance of the top two layers.

Hence, if we replace the two upper layers by a single homogeneous layer with effective thickness h_H and effective resistivity ρ_H, it is easily seen that they are given by:

$$h_H = h_1 + h_2 \tag{3.24}$$

and:

$$\rho_H = \frac{h_1 + h_2}{S_1 + S_2} = (h_1 + h_2)/(h_1/\rho_1 + h_2/\rho_2) = \rho_s \tag{3.25}$$

These are sometimes known as Hummel's parameters.

On double-logarithm paper, the first eq.(3.24) represents a straight line parallel to the ordinate; the second eq.(3.25) represents a straight line inclined at an angle of 45° to the axis, cutting it at $S_1 + S_2$. The point of intersection (Fig. 3.7) of these two straight lines is known as Hummel point H, the coordinates (x_H, y_H) of which determine ρ_H and h_H. Eqs.(3.24) and (3.25) are utilized to plot auxiliary H-point Ebert charts for a large number of values of $\mu_2 = \rho_2/\rho_1$ and different values of $v_2 = h_2/h_1$ with:

$$\text{abscissa} = x_H/h_1 = 1 + v_2$$

and:

$$\text{ordinate} = y_H/\rho_1 = \left(\frac{1 + v_2}{1 + v_2/\mu_2} \right)$$

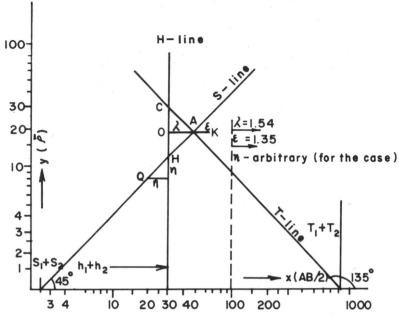

Fig. 3.7 Triangle of anisotropy. Positions of A, K - and Q - points with respect to H - point shown for a particular case: $\rho_1 = 5$ Ωm, $\rho_2 = 40$ Ωm, $h_1 = 10$ m, $h_2 = 20$ m, η - arbitrary.

These auxiliary point charts (H) have been presented in Fig. 3.8(b), which can be directly used to find the H-point (i.e., ρ_H and h_H). The conditions for which the H-point charts have been plotted are strictly valid only for $\rho_3 = \infty$. It is, however, found that the charts can be used even for any arbitrary value of ρ_3 sufficiently greater than ρ_2. This forms the basis of a simple method of construction of empirical curves of the H-type.

3.4.2 Case II: A-type

In this case, the intermediate layer is more resistive than the first (i.e., $\rho_2 > \rho_1$). Hence, the effect of the transverse resistance T cannot be neglected. Also, since $\rho_3 > \rho_2$, we have additionally to take into account the longitudinal conductance S. Therefore, the T and S values of the reduced homogeneous layer will be the sum of those of the top two layers, i.e.:

$$T = T_1 + T_2 = h_1\rho_1 + h_2\rho_2$$

and:

$$S = S_1 + S_2 = h_1/\rho_1 + h_2/\rho_2$$

This case has already been dealt with, and the effective thickness and resistivity are given by - see eqs.(3.22) and (3.23):

AUXILIARY GRAPH (A-Type)
(Ebert Chart)

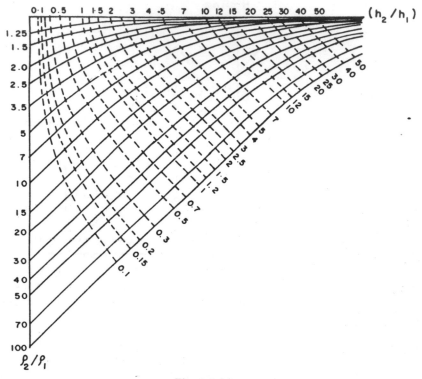

Fig. 3.8 (a)

$$h_A = \sqrt{TS} = \lambda(h_1 + h_3) = \lambda H \qquad (3.26)$$

and:

$$\rho_A = \sqrt{T/S} = \rho_m = \lambda\rho_s \qquad (3.27)$$

These points are obviously the coordinates of the point of intersection of the T and S lines (Fig. 3.7). This point A may be called the "anisotropy point". Thus, the thickness of the reduced layer equivalent to the first two layers is not equal to the sum of the thicknesses (as in case I, i.e., H-type), but is λ times that. The resistivity of the reduced layer is also λ times the longitudinal resistivity of the first two layers.

As in case I, here also eqs. (3.26) and (3.27) can be utilized to plot the auxiliary point charts (A) for various values of μ_2 and v_2. These relations (3.26) and (3.27) may be rewritten as:

$$\frac{x_A}{h_1} = \{(1 + v_2/\mu_2)(1 + v_2\mu_2)\}^{\frac{1}{2}}$$

AUXILIARY GRAPH (H – TYPE)
(Ebert Chart)

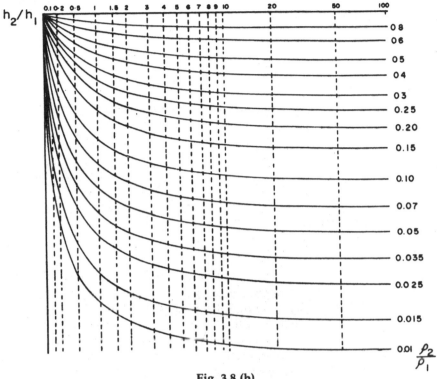

Fig. 3.8 (b)

$$\frac{y_A}{\rho_1} = \left\{ \frac{1 + v_2 \mu_2}{1 + (v_2 / \mu_2)} \right\}^{\frac{1}{2}} \tag{3.28}$$

The values obtained from (3.28) are plotted in a similar manner as in case I, and have been presented in Fig. 3.8(a). The coordinates of the point A (h_A and ρ_A) can be calculated, either by using the above relations (3.28), or can be read from the curves available (Fig. 3.8a).

3.4.3 Case III: K-type (modified A-type)

In this case, the resistivity of the intermediate layer is higher than that of the top and the bottom layers. The current flow in the upper two layers will be somewhat similar to case II, i.e., A-type, especially at smaller AB.

Therefore, both T and S should be considered. In this case, however, the intermediate layer being underlain by a less resistive layer, the lines of current flow within the second layer should have a larger vertical component than that of A-type. Hence, the conditions should be somewhat different than those of A-type section.

AUXILIARY GRAPH (K – Type)
(Ebert Chart)

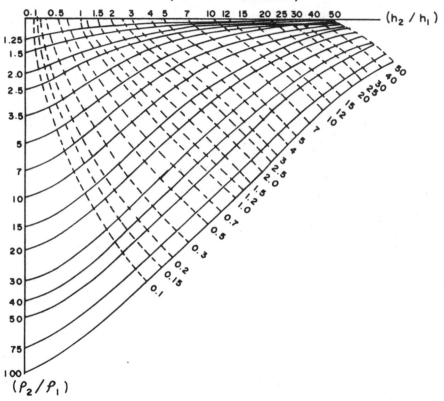

Fig. 3.9(a)

From an examination of the theoretical curves it is found that, in this case, the resistivity of the reduced layer remains the same as in the A-type section whereas its thickness is greater than in the case of the A-type section. Here, h_K is equal to $\varepsilon\lambda(h_1 + h_2)$, where ε is a function of λ always greater than unity. The empirical relationship between λ and ε is shown in Fig. 3.10. The value of ε can be read from this graph for different values of λ.

AUXILIARY GRAPH (Q – TYPE)
(Ebert Chart)

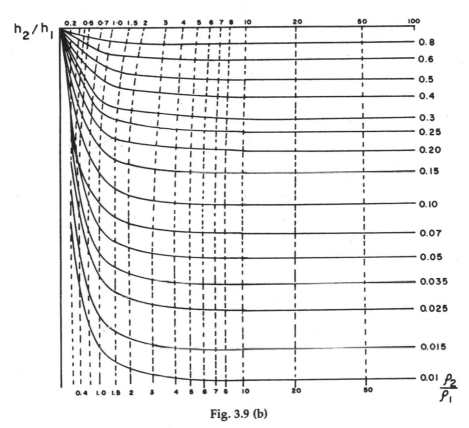

Fig. 3.9 (b)

The values of λ and the corresponding values of ε may be tabulated as given in Table 3.1 (Kalenov, 1957) or graphically presented (Fig. 3.10).

Table 3.1: Variation of ε with Coefficient of Anisotrophy

λ	1.10	1.20	1.30	1.40	1.50	1.70	2.00	2.50	3.00
ε	1.17	1.17	1.29	1.32	1.33	1.36	1.38	1.40	1.42

Thus, the parameter of the reduced layer is determined by the coordinates of the point K, given by.

$$x_K = \varepsilon\sqrt{TS}$$

$$y_K = \sqrt{T/S} \qquad (3.29)$$

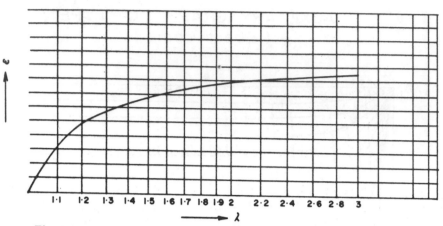

Fig. 3.10   Variation of the factor ε with coefficient of anisotropy (λ).

Point K may be called the "displaced anisotropy" point. On double-logarithm paper, the abscissa of A is $\log[\lambda(h_1 + h_2)]$, and the abscissa of H (i.e., "O") is $\log(h_1 + h_2)$, as shown in the diagram (Fig. 3.7), then:

$$OA = \log\{\lambda(h_1 + h_2)\} - \log(h_1 + h_2) = \log\lambda$$

The triangle HAC is called the triangle of anisotropy, and the height (OA) of it is equal to the coefficient of anisotropy λ. The location of point K is then obviously on the line OA, produced such that $AK = \log\varepsilon$.

The equations for plotting the Ebert chart (K) may be written as:

$$\frac{x_K}{h_1} = \varepsilon\{(1 + v_2/\mu_2)(1 + v_2\mu_2)\}^{\frac{1}{2}} \tag{3.30}$$

$$\frac{y_K}{\rho_1} = \left\{\frac{1 + v_2\mu_2}{1 + v_2/\mu_2}\right\}^{\frac{1}{2}}$$

The curves for various values of v_2 and μ_2 have been presented in Fig. 3.9(a) and the K-point can be located either with the help of those charts or from the above relations (3.30).

3.4.4   Case IV: Q-type (modified H-type)

In this case, the resistivity of the intermediate layer is less than that of the top layer. Hence, the initial part of the sounding curves resembles that of H-type sections; but auxiliary H-point charts cannot be used, as the Q-type is underlain by a less resistive layer.

In this case, it is found empirically that the total thickness of the reduced layer is less than $(h_1 + h_2)$ by a factor η, depending on the values of μ_2 and v_2 of the electrical section. The value of η can be read from the curves given in Fig. 3.11 plotted for various values of μ_2 and v_2.

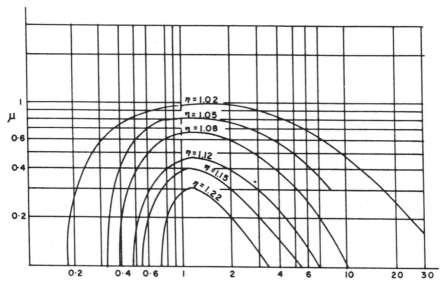

Fig. 3.11 Dependance of the empirical relation between λ and μ on factor η.

The effective resistivity of the reduced layer is also taken as less than the mean longitudinal resistivity by the same factor η. Then the coordinates of Q are given by:

$$x_Q = \frac{H}{\eta}$$

$$y_Q = \frac{1}{\eta}\frac{H}{S} \tag{3.31}$$

It is easily seen that point Q (Fig. 3.7) is graphically found by decreasing η times the coordinates of H. The location of the Q-point with reference to the H-point has been shown in the "triangle of anisotropy" diagram (Fig. 3.7) for a specific value of η.

The set of equations (3.31) may be rewritten to plot the auxiliary point charts(Q) for various values of μ_2 and v_2 as:

$$\frac{x_Q}{h_1} = \frac{1}{\eta}(1 + v_2)$$

$$\frac{y_Q}{\rho_1} = \frac{1}{\eta}\frac{1 + v_2}{1 + (v_2/\mu_2)} \tag{3.32}$$

The Q-point may be taken from the charts (Fig. 3.9b) or calculated from these relations (3.32) with the help of Fig. 3.11.

3.5 Merits and Demerits of the Principle of Equivalence

It has been mentioned earlier (Sec. 2.5, Chapter 2) that for certain relations of the parameters of a three-layer section, changes in resistivity and thickness of the intermediate layer do not produce any noticeable changes in the form of the sounding curve. In such cases it is not possible to distinguish between the different intermediate layers, and error may be involved in the interpretation of such sections.

Thus, for sections of the H or A-type, it is found that the form of the curve remains practically the same if within certain limits the values of h_2 and ρ_2 are multiplied by the same factor. In other words, H or A-type sections are practically equivalent within certain limits if $h_2/\rho_2 = S_2$ remains the same. These sections may be called equivalent with respect to S.

Also, for sections of K or Q-type, the form of the curves does not change appreciably if h_2 is increased or decreased by a certain factor and ρ_2 is correspondingly decreased or increased by the same factor. Thus, if $h_2\,\rho_2 = T_2$ remains constant, any change of h_2 and ρ_2 separately does not produce any noticeable change in the form of the curve. Such sections may be called equivalent with respect to T.

The mathematical proof of this principle has also been given under certain conditions of thickness and resistivity of the intermediate layer (eqs. 2.72 and 2.73 in Chapter 2). Let us examine a few examples of equivalent curves of the four types.

(a) H-types; $\rho_3 = \infty$. In this case, the curve $h_2/h_1 = 1$; $\rho_2/\rho_1 = 1/19$, i.e., $S_2/S_1 = 19$, is almost equivalent to the curve given by: $h_2/h_1 = 2$; $\rho_2/\rho_1 = 1/9$ i.e., $S_2/S_1 = 18$ (refer to sets of charts 84s and 85s in COMPAGNIE GENERALE DE GEOPHYSIQUE, 1955, 1963).

(b) A-type; $\rho_3 = \infty$. Here the equivalent curves are given by:

$$v_2 = 2;\ \mu_2 = 39 \text{ i.e., } S_2/S_1 = 1/19.5 \text{ (refer to set 96s)}$$

and:

$$v_2 = 1;\ \mu_2 = 19 \text{ i.e., } S_2/S_1 = 1/19 \text{ (refer to set 97s)}$$

(c) K-Type; $\rho_3 = 0$. The equivalent sets are given by:

$$v_2 = 2\ ;\ \mu_2 = 19;\ T_2/T_1 = 38 \text{ (refer to set 91s)}$$

and:

$$v_2 = 1;\ \mu_2 = 39;\ T_2/T_1 = 39 \text{(refer to set 92s)}$$

(d) Q-type: $\rho_3 = 0$

$$v_2 = 2;\ \mu_2 = 1/39;\ T_2/T_1 = 1/19.5 \text{ (refer to set 80s)}$$

and:

$$v_2 = 1; \mu_2 = 1/19; T_2/T_1 = 1/19 \text{ (refer to set 79s)}$$

are the equivalent curves.

The principle of equivalence as shown above, applies only for small values of $v_2 = h_2/h_1$. The ratio may be different for different values of $\mu_2 = \rho_2/\rho_1$ and $\mu_3 = \rho_3/\rho_1$. For interpretation of sounding data it is thus important to know the maximum values of h_2/h_1 beyond which the principle of equivalence does not hold good (Table 3.2). Besides, for interpretation of field curves, it is of practical importance to know the limits within which h_2 and ρ_2 may be varied, satisfying the equivalence of the curves.

Table 3.2: Table of Limited Values for v_2 and μ_2 with Respect to S and T

Equivalent with respect to S			Equivalent with respect to T		
Maximum value of v_2 for which μ_2 and v_2 can be decreased without limit	Maximum possible value of factor of increase of v_2 and μ_2		Maximum value of v_2 for which μ_2 can be increased and v_2 decreased without limit	Maximum possible factor of decrease of μ_2 and increase of v_2	
H-type ($\rho_3 = \rho_1$)			K-type ($\rho_3 = \rho_1$)		
μ_2	v_2		μ_2	v_2	
1/39	2	1.6	39	9	1.7
1/19	1	1.6	19	5	1.6
1/9	1	1.6	9	2	1.6
1/4	1/2	1.4	4	1/2	1.5
3/7	1/3	1.4	7/3	1/2	1.5
2/3	1/5	1.4	3/2	1/3	1.5
A-type ($\rho_3 = \infty$)			Q-type ($\rho_3 = 0$)		
39	3	without limit	1/39	–	–
19	2	-do-	1/19	1/3	2.6 - 2.3
9	1	3.8	1/9	1/3	2.2 - 2.1
4	1	2.5	1/4	1/3	1.9 - 1.9
7/3	1	1.7	3/7	1/3	1.8 - 1.8
3/2	1	1.5	2/3	1/3	1.5 - 1.5

Assuming that the error in field measurements is 5%, we may say that the curves having deviation less than 5% cannot be differentiated from each other. Let us consider, for example, the theoretical curves of the K-type having constant T_2/T_1, $\rho_3 = 0$ and variables h_2/h_1 and ρ_2/ρ_1. It is seen that for $\rho_2/\rho_1 > 9$, the curves are equivalent within 5%. In other words, curves for $\rho_2/\rho_1 = 9$, 19 and 39 practically coincide and may be taken as the same as for $\rho_2/\rho_1 = \infty$ and $T_2/T_1 = h_2 \rho_2/h_1 \rho_1 = 12$. Thus, if

$\rho_2/\rho_1 = 9$, then starting with $h_2/h_1 = 1.3$, we can increase ρ_2/ρ_1 and correspondingly decrease h_2/h_1 without limit and shall not notice much difference in the nature of the curves. If $\rho_2/\rho_1 = 19$, the maximum limit of h_2/h_1 will be 0.6, and so on.

The problem of equivalence has been discussed in detail by Bhattacharya and Patra (1968). This includes nomograms for all four types of three-layer sections, which give the limiting values of ρ_2 and h_2 for the principle to be valid. The nomograms are available in Plates V-VIII in Bhattacharya and Patra (1968). From these nomograms we can find the numerical values of the limits of applicability of the principle of equivalence for different types of sections. Some numerical values of the maximum limits of equivalence are given here in Table 3.2.

It is found that the region of equivalence in the case of Q-type section is considerably less than that in the cases of A, K and H-type sections.

The principle of equivalence plays an important role in the graphical construction of field curves as well as in their interpretation. Suppose, for example, it is necessary to find an H-type curve with parameters $\mu_2 = 1/30$, $v_2 = 4$, and $\rho_3 = \infty$. A theoretical curve with such parameters does not exist in the sets of master curves published. The closest theoretical curve has the value $\mu_2 = 1/39$. Pylaev's nomogram (see Bhattacharya and Patra, 1968, p. 63, Plates V-VIII) shows that for the given parameters, μ_2 may be changed within the limits of equivalence. Thus, the corresponding value of v_2 is given by $v_2/\mu_2 = 120 = v_2'/\mu_2'$, i.e., $v_2' = 120/39 = 3.1$. Thus, we can use the curve $\mu_2 = 1/39$, $v_2 = 3$, $\rho_3 = \infty$, which is equivalent to the given section within an error of about 5%. To get a K-type curve, equivalent with respect to T, for the values $\mu_2 = 30$, $v_2 = 4$ and $\rho_3 = 0$ we can use the relation $v_2' = v_2\mu_2/\mu_2'$ and $\mu_2' = v_2\mu_2/v_2'$. Then the parameters of the new curve equivalent to the given K-type section are: $\mu_2 = 39$, $v_2 = 3$ and $\rho_3 = 0$.

It should be remembered that for Q-type and A-type sections the procedure is exactly the same as for those of K and H-type, respectively.

Principle of equivalence imposes a restriction on the interpretation of geosounding data by introducing error in precise determination of thicknesses small compared to depth. The ambiguity is to be removed through geological and other controls.

We face difficulty in detecting beds having resistivity values intermediate to those of the enclosing beds. If the bed is not considerably thick its effect will not be reflected on the apparent resistivity curve. When the thickness of the bed increases it has to compensate the effect due to the increase in resistivity of the enclosing beds, controlled by the "principle of suppression".

3.6 Interpretation Methods

The interpretation of Vertical Electrical Sounding (VES) data obtained from field surveys is based on an analysis of Kernel function controlling the potential expression on the surface of a horizontally stratified earth (eq. 2.53). The general expression for potential over a layered earth may be written as:

$$V = \frac{I\rho_1}{2\pi} \left[\frac{1}{r} + 2 \int_0^\infty A(m) J_0 (mr) dm \right] \qquad (3.33)$$

Using Weber's integral formula eq. (3.33) can be written as:

$$V = \frac{I\rho_1}{2\pi} [1 + 2A(m)] J_0 (mr) dm \qquad (3.34)$$

where $A(m)$ is the Kernel of the integral controlled by the values of the layer parameters for the horizontally stratified earth. The evaluation of the Kernel function is shown in Sec. 2.4.2 for a two-layer earth and in Sec. 2.4.3 for a three-layer earth.

The Kernel function $A(m)$ is referred to as Stefanescu (also Stefanesque or Stafanesco) Kernel. Other related useful functions are as follows :

Koefoed's raised Kernel function, $H(m)$

Koefoed's modified Kernel function, $G(m) = \dfrac{H(m) - \dfrac{1}{2}}{H(m) + \dfrac{1}{2}}$

Slichter's Kernel function, $K(m) = [1 + 2A(m)]$
Resistivity transform, $T(m) = \rho_1 K(m) = \rho_1 [1 + 2A(m)]$

Another important relationship between resistivity transform, $T(m)$ and apparent resistivity, $\bar{\rho}$ is quite useful in the interpretation and is given as follows:
Eq.(3.34) gives,

$$V = \frac{I}{2\pi} \int_0^\infty \rho_1 [1 + 2A(m)] J_0 (mr) dm = \frac{I}{2\pi} \int_0^\infty T(m) J_0 (mr) dm \qquad (3.35)$$

and :

$$E = \frac{I}{2\pi} \int_0^\infty T(m) J_1 (mr) m dm \qquad (3.36)$$

and :

$$\bar{\rho} = 2\pi r^2 (E / I) = r^2 \int_0^\infty T(m) J_1 (mr) m dm \qquad (3.37)$$

The values of $A(m)$ for two-layer and three-layer cases are given in Sec. 2.4.2 and Sec. 2.4.3 respectively, which may be used for computation of theoretical master curves through forward solution for vertical electrical sounding in a series calculation (eqs. 2.59 and 2.69). These master curves are used for interpretation of layer parameters from VES curves which forms the basis of indirect approach for interpretation.

3.6.1   Indirect Approach

The indirect approach for interpretation of VES curve is to generate theoretical curves and compare with field curves until a match is obtained. Known values of layer parameters are fed through the Kernel and curve is generated. Instead of the computation of a long series, Ghosh (1970, 1971b) introduced a rapid inverse filter method based on a special case of Watson's transform. This replaces the integral by a convolution process and does not place any restriction as to the total number of layers present or to their thickness.

Ghosh's inverse filter method is a fast method of computing apparent resistivity ($\bar{\rho}$) curves for known layer parameters based on the application of a linear filter through the concept of $T(m)$ (eq. 3.37) related to Stefanescu Kernel (eq. 3.35) as:

$$V = \frac{I}{2\pi} \int_0^\infty T(m) J_0\ (mr)dm$$

The forward problem using Ghosh's inverse filter coefficients will be detailed in Chapter 4. This will form the basis of the conventional "curve matching" technique of interpretation of VES data.

3.6.2   Direct Approach

The method used to evaluate layer parameters directly from the VES curves without using the conventional manual "Curve matching" is referred to as 'direct' one against the background of indirect approach outlined in Sec. 3.6.1. Following Slichter (1933) and Pekeris (1940), Koefoed (1968) gave the first detailed account of 'direct interpretation of VES curves' wherein he introduced the concept of "raised kernel" function, $H(m)$ and "modified kernel" function, $G\ (m)$, to derive parameters. Ghosh (1971a) devised a simple and rapid procedure for getting the value of resistivity transform, $T(m)$ related to $A(m)$ and subsequently layer parameters using Ghosh's filter method (Chapter 4).

3.7   Schlumberger Curve Matching and Ebert Charts

The theoretical derivations (Bhattacharya and Patra, 1968) — given earlier form the basis of "auxiliary point method" of interpretation of Schlumberger sounding with the help of album of standard two-layer theoretical curves and Ebert charts. The results of such interpretation, referred to as preliminary interpretation, may give us a fairly accurate representation of layer parameters under favourable circumstances with suitable geological control. However, this being a grapho-analytical approach has its intrinsic drawbacks and therefore, may not reliably obtain the correct geoelectric parameters with a prescribed degree of exactitude and precision necessary for accurate assessment of the data. In order to check the validity of the results of preliminary interpretation, these are fed into the computer and synthetic sounding curves obtained. A comparison is made between the two curves and final interpretation achieved through trial and error, if necessary.

The curve matching techniques for interpretation of Schlumberger VES curves using two-layer master curves (Fig. 3.1) and Ebert charts (Figs. 3.8 and 3.9) are given as follows:

(a) Each of the branches of an apparent resistivity curve is approximated by a two-layer one.

(b) The coordinates of the cross of this two-layer curve are considered to represent the thickness and the resistivity of a fictitious layer that replaces the sequence of shallower layers.

(c) To get the parameters for the fictitious layer, sets of graphs are used. The coordinates on these graphs are the ratio of the thickness of the replacing layer to that of the first layer, and the ratio of the resistivity of the replacing layer to that of the first layer. The parameters are thickness ratio and resistivity ratio plotted on double-log graphsheet of modulus 62.5 mm.

(d) Four sets of such auxiliary point charts are available for H, A, K and Q types (Figs. 3.8 and 3.9).

3.7.1 Interpretation of Two-layer Curve

The indirect approach briefly outlined in Sec. 3.6.1 comprises curve matching technique for a preliminary interpretation of layer parameters manually. These layer parameters are also needed as initial guess parameters in the inversion of resistivity data (Sec. 3.6.2).

Figures 3.4 and 3.5 are self-explanatory for the interpretation of two-layer field curves by curve matching. The method adopted for finding ρ_1 and h_1 has been indicated in Fig. 3.5. The procedure is to plot the field curve on a double-log transparent graphsheet of modulus 62.5 mm (same as that for the theoretical master curves), and then to superpose the transparent graphsheet on the master curve (Fig. 3.1a). The transparent graphsheet containing field curve is then moved over the master curve keeping the axes parallel (Fig. 3.4). Figure 3.5 shows the matched condition, and the point on the transparent double-log sheet coinciding with the origin of master curve, gives h_1 along the abscissa and ρ_1 along the ordinate. The values of h_1 and ρ_1 are 8m and 4 ohm-m in the present case (Fig. 3.5).

3.7.2 Interpolation on Log Scale

Interpretation of field curves by curve matching technique needs sufficient knowledge of interpolation on logarithmic scale as the field curve is likely to lie between curves in the master set. This is explained through Fig. 3.12 for two cases: (i) when the field curve lies between two master curves (Fig. 3.12a), and (ii) when it lies outside two master curves (Fig. 3.12b). The solid lines represent the field curves, and dashed lines the master curves.

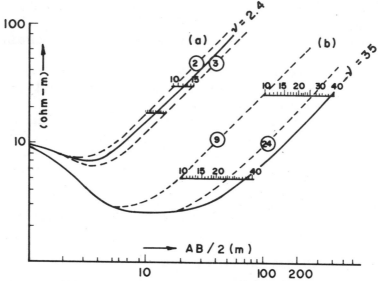

Fig. 3.12 Method of reading by interpolation on logarithm scale.
(a) Field curve lying between two master curves, interpolation.
(b) Field curves lying out side two master curves, extrapolation.
Beginning of a cycle coincided with the left hand master curve.
Unbroken curve = field curve; dashed curve = three-layer master curve.

The procedure is to cut out a piece of logarithm paper of the same modulus (62.5 mm) and superpose this on the divergent or parallel part of the curves. The same scale divisions as the difference in v_2 values between the two master curves within or without which the field curve lies should be included if possible. A paper with a logarithm scale and modulus of 62.5 mm may be used conveniently. If no such scale is available, one can be prepared by pasting cut-out portions of a graph sheet onto any paper scale. The beginning of a cycle must coincide with one of the master curves in question, as shown in the diagram. In Fig. 3.12(a), the field curve lies between three-layer master curves for $v_2 = 2$ and 3. The interpolated value of v_2 for the field curve is found to be equal to 2.4. In Fig. 3.12(b), the field curve is outside $v_2 = 9$ and 24. The extrapolated value of v_2 is found to be equal to 35.

3.7.3 Interpretation of a Four-layer HK-type Field Curve

The procedure of interpretation of a four-layer *HK*-type curve (referred to as working graph) with the help of two-layer master curves (Fig. 3.1) and corresponding Ebert Charts (Figs. 3.8 and 3.9) may be illustrated as follows:

(i) The four-layer field curve comprises of three-layer branches.
(ii) Superimpose the working graph (solid line) on a family of two-layer master curves (Fig. 3.1b) and approximate the first branch of the field curve. The origin of the master set read on the field curve at the matched condition is the "first cross" (+) giving h_1 and ρ_1 values. Read ρ_2/ρ_1 from the dashed curve. Here, $\rho_1 = 16$ ohm-*m*, $h_1 = 1.3$ m and $\rho_2/\rho_1 = 1/3$ (Fig. 3.13).

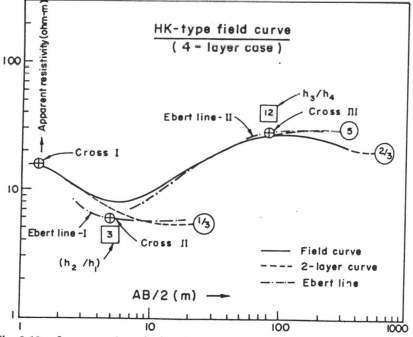

Fig. 3.13   Interpretation of a four-layer *HK* - type curve using two-layer master curves and Ebert charts.

(iii) The working graph is then superimposed on H-type Ebert charts (Fig. 3.8b) with its origin on the "first cross" and corresponding Ebert line I (dash-dot) is copied for $\rho_2/\rho_1 = 1/3$.

(iv) Superimpose the working graph on two-layer set (Fig. 3.1a) and approximate to the second part of the curve (dashed line) keeping master set-origin on the Ebert line I. Draw the "second cross" (+) as read on working graph. The values read on "second cross" gives ρ_H and h_H and the resistivity ratio ρ_3/ρ_H. The values here are : $\rho_H = 6$ ohm-m, $h_H = 5\ m$ and $\rho_3/\rho_H = 5$.

(v) The step (iii) is repeated and Ebert line II is drawn keeping "second cross" at resistivity ratio '5' on the ordinate of K-type Ebert chart (Fig. 3.9a).

(vi) Finally, approximate the last branch of the curve by the two-layer master set (Fig. 3.1b) keeping its origin on Ebert line II and noting ρ_4/ρ_K. The origin read on the working graph, the "third cross" gives ρ_K and h_K values. Last two-layer match gives ρ_4/ρ_K equal to 2/3. Here, $\rho_K = 29$ ohm-m and $h_K = 80\ m$. Thickness ratios h_2/h_1 and h_3/h_H as read w.r.t. cross I and cross II are 3 and 12, respectively.

(vii) Final interpreted layer parameters are as follows (Fig. 3.13) :
$h_1 = 1.3\ m$, $h_2 = 3.9\ m$, $h_3 = 60\ m$, $h_4 = \infty$
$\rho_1 = 16$ ohm-m, $\rho_2 = 12$ ohm-m, $\rho_3 = 30$ ohm-m, $\rho_4 = 20$ ohm-m.

3.7.4 Interpretation of Multi-layer HKQ-type Field Curve

The procedure adopted here, is briefly outlined below:

(a) Approximate the first part of the apparent resistivity field curve by a two-layer descending type master curve. The Cross-I read over the origin gives first layer thickness and resistivity.

(b) Superpose the curve on H-type Ebert chart with origin over the origin and draw Ebert line-I corresponding to resistivity ratio obtained in step(a).

(c) Next superpose the field curve over a two-layer ascending type master curve and approximate to the second branch taking care that origin lies on Ebert line-I and record this origin as Cross-II. This gives third layer resistivity with reference to Cross-II. Read Cross-II w.r.t. Cross-I to note second layer thickness w.r.t. Cross-I.

(d) Superpose the field curve upon K-type Ebert chart with Cross-II on its vertical axis corresponding to resistivity ratio in step(c). Copy this parameter line on the field curve to get Ebert line-II.

(e) Superpose the field curve on two-layer descending type master curve and note the resistivity ratio with cross-III on Ebert line-II. Read the thickness ratio on K-type Ebert-chart for Cross-III w.r.t. Cross-II.

(f) The last part of the curve represents descending type. Therefore, superimpose the field curve on a two-layer descending type master curve taking care the cross-IV lies over Ebert line-III, drawn with Q-type Ebert chart.

(g) The two layer curve that gives the best fit finally defines the last resistivity ratio. Read the thickness ratio from Cross-IV with reference to Cross-III on Q-type Ebert chart.

Although we depend much on computer for interpretation of VES data, curve matching is an essential tool for getting the layer parameters easily, particularly as an initial guess for computer-aided interpretation (Chapter 4).

3.8 Effect of Dip on Interpretation

We have so far discussed the apparent resistivity curves only for a horizontally stratified earth, and the interpretation techniques explained in previous sections apply strictly when the boundaries between different layers are horizontal. Often, vertical electrical sounding is carried out in regions where the boundaries of separation between different layers are inclined. It is necessary, therefore, to examine the effect of the inclination of beds on VES curves.

We reproduce here two sets of curves (Fig. 3.14a, b) from

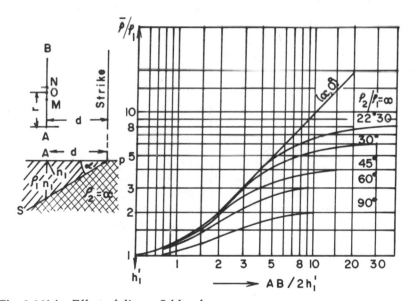

Fig. 3.14(a)   Effect of dip on Schlumberger curves.
(a) Expansion of electrodes parallel to strike. Lower layer highly resistive.

Bhattacharya and Patra (1968) to show the effect of dip on VES curves for soundings parallel to strike (Fig. 3.14a) and perpendicular to the strike (Fig. 3.14b). In Fig. 3.14(a), the apparent resistivity in terms of first layer resistivity (ordinate) is plotted against $AB/2h_1'$ (abscissa), where h_1' is the

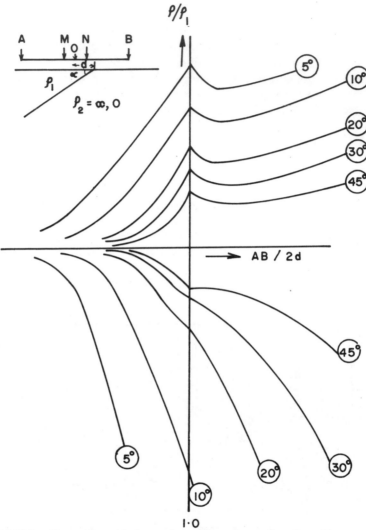

Fig. 3.14(b)   Expansion of electrodes across the strike; d = distance of the sounding point from the contact.

depth perpendicular (not the vertical depth, h_1) to the boundary. The curves have been plotted for values of the angle of inclination (dip) $\alpha = 0°$, 22°30′, 30°, 45°, 60° and 90°, with an infinite resistivity contrast (i.e., $\rho_2/\rho_1 = \infty$). Similar curves are also available for different values of resistivity contrast.

Master curves are also available for inclined contacts with spread of electrodes perpendicular to the strike direction (Fig. 3.14b). With the increase of the angle of inclination the curves, here, differ widely from each other such that an increase in α raises the left hand part of the curve and lowers the right hand part. If the two-layer master curves (Fig. 3.1) are used to interpret the sounding curves, for an inclined two-layer contact, considerable error is involved, particularly when α is large.

However, if the angle of inclination (α) is not more than 15-20°, the ordinary two-layer master curves may be used without appreciable error (less than ten per cent). If the basement must be surveyed when α is small, it is obvious that the Schlumberger sounding is unsuitable. In such a case, dipole sounding may be carried out.

In practice, if it is known before hand that the boundary is inclined, then quantitative interpretation is possible when one of the parameters (α or ρ_2/ρ_1) is known. Then, using the master curves for inclined contact, the other parameter and the depth of the bottom layer can be determined. In an unknown area, the dip of the bed can be detected by carrying out sounding at a point in two mutually perpendicular directions. If the strike direction is approximately known, the sounding should be done at several points, with a spread parallel to the strike direction and the depths computed from the two-layer master curves for horizontal layers. Thus, the trace of the lower bed and the angle of inclination can be found. The interpretation is reasonably accurate for small dips. For larger dips, however, the master curves for dipping contacts should be used.

4 Computer Aided Interpretation

4.1 General Consideration

Slichter (1933) used the solution by Langer (1933) to interpret the resistivity of a horizontally layered earth model from the known surface potentials. Vozoff (1958) was the first to use the steepest descent method for resistivity data. Zohdy (1965, 1974) has developed a method of direct automatic interpretation using modified Dar Zarrouk parameters. Meinardus (1970) used a sum of squares of difference between sample values of Slichter Kernel function obtained from field data and response of the model. Johansen (1977) considered the minimization of sum of squares of logarithm of resistivity and thickness of layers and the theoretical responses.

Backus and Gilbert (1967, 1968, and 1970) showed that generalized inversion works very well on the geophysical problems. However, the solution is somehow dependent on the initial guess. Marquardt (1970) has proposed the ridge regression technique and showed how near singular data could be refined by using biased estimators. He showed that ridge regression is preferable for the problems with small eigenvalues while generalized inversion is preferable for problems with some zero eigenvalues.

Inman et al. (1973) were the first to apply the generalized inversion method to resistivity problem. Inman (1975) has found that the resistivity problems involve small but rarely zero eigenvalues and, hence, the ridge regression method is suitable for inversion of resistivity data.

Hoverstein et al. (1982) by studying different least square inversion techniques concluded that ridge regression is the fastest.

In the Inverse theory, we encounter two types of problem: (i) large model space and (ii) multimodal error function. Existing calculus based enumerative methods are local in scope and get easily trapped in local minima of the energy function. Non-linear least square inversion methods came into existence when direct methods failed. They have been used based on Tarantola's formulation. The essential requirement of this method is good starting model since it looks for a solution in the close vicinity of the starting model.

Recently, a new class of methods to solve non-linear optimization problem has generated considerable interest, in the field of artificial intelligence. These methods are able to solve highly non-linear and non-local optimization problems and belong to the class of global optimization techniques which includes simulated annealing, genetic algorithm and evolutionary programming along with neural network and more advanced fuzzified neural network systems.

With the advent of high speed computers having parallel processing environment, the use of global optimization methods in search for higher resolution in resistivity parameters for the systematic evaluation of aquifer properties has become inevitable.

4.2 The Forward Problem

Numerical modelling is the term used to represent an approach in which a true earth structure is replaced by one for which numerical approximation of Poisson's equation can be made and evaluated. In the decade of the 1970s, methods for numerical and analytical modelling of the interaction of electric fields with the earth structure developed rapidly. This development was driven by the availability of highly capable computers with which such models can be constructed. This modelling capability has made possible the extraction of much more information from field data than had been possible previously. The modern capabilities of geophysical methods is based on two technical developments: the ability to acquire great volume of data with high accuracy, and the possibility of extracting sophisticated models of the geoelectric structure from these data. A brief description of the forward problem has been given earlier in Sec. 3.6.1 under indirect approach.

4. 2.1 The Importance of the Forward Model

Interpretation requires modelling to know if the interpretation put forth is compatible with the data. The range of models that can be put forth is so great that some automated scheme of producing possible solutions is needed. Finding models that satisfactorily predict the observed data is called the inverse problem. The inverse problem involves predicting the model given the data. The inverse problem has been briefly outlined in Sec. 3.6.2 under direct approach.

All the classes of inverse problem involve a good deal of computation time. For each iteration of inversion the forward model is evaluated and it's response is compared in the inverse algorithm. For a non-linear problem the optimization process requires a large number of iteration and hence the forward model is calculated a large number of times.

This puts a heavy premium on efficient ways of solving the forward problem. There are two other factors that must also be considered when dealing with modelling methods for resistivity inversion. The first is the need to model a very large medium since the conductivity structure are generally not known beforehand. The other factor is the need to deal with the regional problem.

4.2.2 Various Approaches to Solve the Forward Problem

There are five common approaches to solve the forward problem given below:
1. Analytical method
2. Boundary integral or volume integral method
3. Fourier methods
4. Finite element and Finite difference methods and
5. Hybrid approach.

4.2.2.1 Analytical Method

Before the availability of modern high speed computer, an important means of modelling was through the use of mathematical analysis to arrive at the exact solutions of Poisson's equation, which could be evaluated by relatively straightforward numerical techniques. The general solution of Poisson's equation along with a sufficient number of additional conditions can be used to construct special solutions which represent geoelectric structure of real interest. Analytical methods are restricted to geometries in which Poisson's equation is separable. This usually restricts one to the horizontally layered media. When analytical methods are applicable they are usually the best ones.

4.2.2.2 Boundary Integral or Volume Integral Method

Boundary integral or volume integral method is probably the next most efficient method in situations where the anomalous regions define only a few boundaries. An anomalous region buried in an otherwise layered media is an example of a modelling problem that is well suited to these methods. However, as the number of boundary points increase, the efficiency of the method decreases rapidly. In an inverse problem one usually wants to be able to model a fully inhomogeneous media and this will probably make the boundary integral method inappropriate.

4.2.2.3 Fourier Integral, Finite Element and Finite Difference Methods

The three remaining methods are quite closely related except of course that the Fourier methods operate in the Fourier domain instead of the space domain. All the three methods are able to deal with fully inhomogeneous models. One advantage of the Fourier method is that it can be based on existing and well tested frequency domain computer programs. Dealing with regional problems favours methods that can incorporate graded spacing which allow more distant regions to influence the solution. In the case of the Fourier methods a graded spacing could be transformed into uniform spacing before applying the method. The choice of method, therefore, comes down to matters of speed, accuracy and simplicity. The finite difference and the finite element methods lead to large systems of equations to be solved even for simple models. Recent advances in iterative relaxation techniques and continuously increasing computing power will, in the future, allow for finite difference and finite element modelling of more complicated 3D earth models than at the present time.

In developing algorithms for 3D resistivity modelling, we desire to model arbitrarily complex media, which is especially important when inversions are to be done. The advantage of finite difference and finite element methods over integral equation solution lies in the modelling of complex media. Thus, for a given spacing the finite element methods are usually the most accurate, and finite difference methods are the quickest and the simplest.

4.2.3 Resistivity Forward Problem

The general expression for potential over a layered earth can be given as:

$$V = \frac{I\rho_1}{2\pi}\left[\frac{1}{r} + 2\int_0^\infty A(m)J_0(mr)dm\right] \tag{4.1}$$

$$= \frac{I\rho_1}{2\pi}\int_0^\infty [1 + 2A(m)]J_0(mr)dm$$

where, $A(m)$ = Stefanescu Kernel (Stefanescu and Schlumberger, 1930) function relation of which with other functions is given in Chapter 3 (Sec. 3.6).
Thus :

$$V = \frac{I}{2\pi}\int_0^\infty \rho_1 (1 + 2A(m))J_0(mr)dm = \frac{I}{2\pi}\int_0^\infty T(m)J_0(mr)dm \tag{4.2}$$

$$E = \frac{I}{2\pi}\int_0^\infty T(m)J_1(mr)mdm \tag{4.3}$$

$$\bar{\rho} = 2\pi r^2 \frac{E}{I} = r^2\int_0^\infty T(m)J_1(mr)mdm \tag{4.4}$$

For two-layer earth the value of $A(m)$ is given by $A(m) = \dfrac{K_{12}e^{-2mH_1}}{1 - K_{12}e^{-2mH_1}}$,

with usual notations.
For three-layer earth, the value is :

$$A(m) = \frac{K_{12}e^{-2mH_1} + K_{23}e^{-2mH_2}}{1 - K_{12}e^{-2mH_1} - K_{23}e^{-2mH_2} + K_{12}K_{23}e^{-2m(H_2 - H_1)}} \qquad (4.5)$$

Using relations $H_1 = p_1 H_0$, $H_2 = p_2 H_0$ etc. and $e^{-2mH_0} = g$, for three-layer earth we get:

$$A(m) = \frac{K_{12}g^{p_1} + K_{23}g^{p_2}}{1 - K_{12}g^{p_1} - K_{23}g^{p_2} + K_{12}K_{23}g^{(p_2 - p_1)}}$$

As p_1 and p_2 are whole numbers, K_{12} etc. constants, and $A(m)$ is a rational function of g, we can write:

$$A(m) = b_1 g + b_2 g^2 + b_3 g^3 + \dots = \sum_{n=1}^{\infty} b_n g^n \qquad (4.6)$$

Comparing eqs. (4.5) and (4.6):

$$K_{12}g^{p_1} + K_{23}g^{p_2} = [1 - K_{12}g^{p_1} - K_{23}g^{p_2} + K_{12}K_{23}g^{(p_2 - p_1)}]\sum_{n=1}^{\infty} b_n g^n$$

This identity requires that coefficients of any order of g must be identically equal on both sides. As the highest order on the left hand side is g^{p_2}, the coefficient of g of any order greater than p_2 on the right hand side must be zero. The coefficients of $g^{p_2 + t}$, where t is a positive integer, is given by:

$$b_{p_2 + t} - K_{12}b_{p_2 - p_1 + t} - K_{23}b_t + K_{12}K_{23}b_{p_1 + t} = 0$$

and we get the recurrence formula:

$$b_{p_2 + t} = K_{12}b_{p_2 - p_1 + t} + K_{23}b_t - K_{12}K_{23}b_{p_1 + t} \qquad (4.7)$$

Thus, $b_{p_2 + t}$ may be calculated, knowing the value of $b_{p_2 - p_1 + t}$, b_t and $b_{p_1 + t}$. The coefficients upto the maximum value b_{p_2} may be determined from the equation following eq. (4.6). The rest of the coefficients can be determined through eq. (4.7).

The potential can now be written in the form (eqs. 2.67 and 2.68):

$$V = \frac{I\rho_1}{2\pi}\left[\frac{1}{r} + 2\sum_{n=1}^{\infty}\frac{b_n}{\{r^2 + (2nH_0)^2\}^{\frac{1}{2}}}\right]$$

and

$$E = \frac{I\rho_1}{2\pi}\left[\frac{1}{r^2} + 2\sum_{n=1}^{\infty}\frac{b_n r}{\{r^2 + (2nH_0)^2\}^{\frac{3}{2}}}\right]$$

For Schlumberger symmetrical arrangement

$$\bar{\rho} = \rho_1\left[1 + 2\sum_{n=1}^{\infty}\frac{b_n r^3}{\{r^2 + (2nH_0)^2\}^{\frac{3}{2}}}\right]$$

In practice, the thickness of the second and the subsequent layers are usually expressed in terms of the thickness of the first layer, i.e., $H_0 = h_1$. Then:

$$\bar{\rho} = \rho_1\left[1 + 2\sum_{n=1}^{\infty}\frac{b_n \delta^3}{(\delta^2 + 4n^2)^{\frac{3}{2}}}\right] \tag{4.8}$$

where $r/h_1 = \delta$ (eq. 2.69).

Eq. (4.8) forms the basis for computation of geosounding curves.

In a similar approach (Mooney et al., 1966) b_n values are calculated within a program through the recurrence formula (Flathe, 1955) written in a polynomial form. Existing methods (CGG, 1955, 1963; Flathe, 1955; Mooney et al.,1966) depend upon the evaluation of integral of the form given in eq. (4.1). The integral being a product of a Kernel function and a Bessel function, cannot be expressed in terms of elementary functions. A restriction normally imposed for a rapid convergence of the series is that the thickness of individual layers should be multiples of some common thickness, preferably the thickness of the top layer.

We shall, therefore, briefly outline a method given by Ghosh (1970, 1971 b) known as Ghosh's inverse filter method which is based on a special case of the Watson transform. This method replaces the integral by a convolution process and does not place any restriction as to the number of layers present or to their thickness.

4.3 Ghosh's Inverse Filter Method

This is a fast method of computing $\bar{\rho}$ curves for known layer parameters based on the application of a linear filter through the concept of $T(m)$ which is related to Stefanescu Kernel.

Eq. (4.1) can be written as:

$$V = \frac{I}{2\pi} \int_0^\infty T(m) J_0 (mr) dm \qquad (4.9)$$

The following steps are to be followed :

(i) Use the values of known layer parameters to calculate $T(m)$ and draw $T(m)$ curve through

(a) 2-layer formula : $T(m) = \rho_1(1 + 2A(m))$

and $\qquad T_{12}(m) = \rho_1 \dfrac{1 + K_{12} e^{-2mh_1}}{1 - K_{12} e^{-2mh_1}}$

(b) Pekeris (1940) relation may be used to calculate $T(m)$ for a multi-layer earth given by:

$$T_i = \frac{T_{i+1} + \rho_i \tanh(mh_i)}{1 + T_{i+1} \tanh(mh_i)/\rho_i} \qquad (4.10)$$

at different sample points.

Thus, calculation of $T(m)$ forms the *first step* in the determination of $\bar{\rho}$. Draw $T(m)$ curve. The resistivity transform function shows the same asymptotic behaviour as the apparent resistivity function, both for small and large abscissa values. In general, the effect of an increase of $(1/m)$ (m-reciprocal of length) on $T(m)$ is similar to that of an increase of the electrode spacing on $\bar{\rho}$; both of them correspond with an increase of the depth from which information is obtained. Only difference lies in the slope for descending type curves. Plot of (T/ρ_1) vs $(1/mh_1)$ for a two-layer earth looks similar to two-layer apparent resistivity curves, $(\bar{\rho}/\rho_1)$ vs. $(AB/2h_1)$.

(ii) Second step consists of transform sampled values being convolved with Ghosh's inverse filter coefficients:

(a) Take sampled values of $T(m)$

(b) Take Ghosh's inverse filter coefficients (Table 4.1)

Table 4.1: Schlumberger inverse nine point filter (Ghosh, 1970, p.105)

a_{-3}	0.0225	a_0	0.1854	a_3	0.4018
a_{-2}	−0.0499	a_1	1.9720	a_4	−0.0814
a_{-1}	0.1064	a_2	−1.5716	a_5	0.0148

(c) Convolve to get $\bar{\rho}$ using the formula:

$$\bar{\rho}_K = \sum_{j=-3}^{5} a_j T_{k-j} \; ; k = 0, 1, \ldots, 6 \tag{4.11}$$

where k is the sample point.

(d) Calculate $\bar{\rho}$ at each sample point (k) and plot the curve against r (equivalent to $1/m$).

The filter coefficients (Table 4.1) with zero or negative subscripts must be multiplied by the samples of the input function that lie to the right of the point of the output function; these filter coefficients with positive subscripts must be multiplied by the samples of the input function that precede the point of the output function; hence, these filter coefficients are called the 'memory' coefficients. Equations (4.10) and (4.11) together with the coefficients in Table 4.1 may be used for writing a program for calculation of $\bar{\rho}$ over a multi-layer earth.

4.4 Direct Interpretation

4.4.1 The Inverse Problem

Koefoed (1968) gave the first detailed account of the method of so called direct interpretation of resistivity data wherein he introduced the concept of 'raised kernel' function to derive layer parameters. He starts with the expression

$$\bar{\rho} = \rho_1 \left[1 + 2r^2 \int_0^\infty A(m) J_1 \, (mr) m \, dm \right] \tag{4.12}$$

where interpretation consists of two steps:

(i) 'Raised Kernel' function $H(m)$ is derived from apparent resistivity curve given by:

$$H(m) = A(m) + \frac{1}{2}$$

(ii) Layer parameters are derived from $H(m)$.

Koefoed (1968, p.16) introduced decomposition method wherein apparent resistivity function is expressed in partial functions for which corresponding "partial kernel functions" are calculated. He derived the relation:

$$A(m) + \frac{1}{2} = H(m) = \sum_i \Delta_i A(m) \tag{4.13}$$

The steps to be followed are given by Koefoed (1968, p.17) under several conditions. He has presented standard curves on double-log paper for determining the partial kernel functions graphically from $\bar{\rho}$ functions.

In the second step, Koefoed introduced a "modified kernel function" $G(m)$ derived from $H(m)$ such that :

$$G(m) = \left[H(m) - \frac{1}{2} \right] \Big/ \left[H(m) + \frac{1}{2} \right]$$

For large values of m the function has an asymptotic value

$$G'(m) = K_{12} e^{-2mh_1}$$

which can be used to get values of K_{12} and h_1 by fitting the first part of the modified kernel graph to a function of the form Ke^{-2mh}. A match with standard curve (Plate VIII, Koefoed, 1968, p.70) already available ($1/mh$ as abscissa and e^{-2mh} as ordinate) gives h_1 from horizontal shift and k_{12} from vertical shift. Subsequently, the top layer is stripped off and the measurements are carried out at the upper boundary plane of the second layer and the layer parameters derived from original $G(m)$ curve. For multi-layer case, the procedure is repeated until we get a two-layer modified kernel curve.

This graphical method (Koefoed, 1968) is only of academic interest as computer based solutions are now preferred. But the credit for revival of the interest in direct computational method must be given to Koefoed who took into consideration the earlier suggestions by Slichter (1933), Pekeris (1940) and ultimately succeeded in deriving layer parameters directly from measured sounding curve (Koefoed, 1968).

Here, $H(m)$ was used as an intermediate step. The first step of his method (derivation of $H(m)$ from $\bar{\rho}$) is lengthy. Second step was later modified by Koefoed (1970) to obtain layer parameters from $T(m)$.

4.4.2 Ghosh's Filter Method

Although Koefoed (1970) improved the second step (layer parameters from $T(m)$) the first step ($T(m)$ from $\bar{\rho}$ curve) was also quite lengthy. In order to avoid this disadvantage, Ghosh (1971a) derived a simple and rapid procedure for getting $T(m)$ from the observed curve. Ghosh's method is based on the proven assumption that $\bar{\rho}$ and $T(m)$ are linearly related and that the principle of linear electric filter theory may be applied to derive one from the other.

The potential at a point on the surface is given by eq.(4.1) from which we get:

$$\bar{\rho} = r^2 \int_0^\infty T(m) J_1(mr) m \, dm \qquad (4.14)$$

Applying Hankel's inversion transformation of the Fourier-Bessel integral in eq. (4.14), we get for Schlumberger arrangement:

$$T(m) = \int_0^\infty [\overline{\rho}(r) J_1 (mr)/r] \, dr$$

Introducing new variables : $[x = In \, r; \, y = In(1/m)]$

$$T(y) = \int_{-\infty}^\infty \overline{\rho}(x) J_1 (1/e^{y-x}) dx \qquad (4.15)$$

This represents a convolution integral relating input $\overline{\rho}(x)$ and output $T(y)$. The filter characteristics of the operation defined by (4.15) is obtained by taking Fourier transform of the functions and thereby converting them into the frequency domain. This helps in a rapid computation of T from $\overline{\rho}$ field curve through sampling and filter theory. Here, $\overline{\rho}$ function is sampled according to Nyquist rule and the sample values replaced by interpolating sinc function of equivalent peak height and period determined by the sample rate. In view of the special property of the sinc function that it is equal to unity at the sample point considered and zero at all other sample points, addition of a finite number enables us to reconstruct the signal. In practice, instead of a continuous sinc function curve, digital filter method is used. The problem of sampling and the determination of filter coefficient has been discussed by Ghosh (1970). He has given coefficients (Ghosh 1971a, p.204) for Schlumberger 'twelve point' long and 'nine point' short filters (Tables 4.2 and 4.3) a convolution of which with the $\overline{\rho}$ sampled values yields $T(m)$.

Table 4.2 (Nine point filter)		Table 4.3 (Twelve point filter)	
$a_{-2} \rightarrow -0.0723$	$a_3 \rightarrow 0.0358$	$a_{-3} \rightarrow 0.0060$	$a_3 \rightarrow 0.0358$
$a_{-1} \rightarrow 0.3999$	$a_4 \rightarrow 0.0198$	$a_{-2} \rightarrow -0.0783$	$a_4 \rightarrow 0.0198$
$a_0 \rightarrow 0.3492$	$a_5 \rightarrow 0.0067$	$a_{-1} \rightarrow 0.3999$	$a_5 \rightarrow 0.0067$
$a_1 \rightarrow 0.1675$	$a_6 \rightarrow 0.0076$	$a_0 \rightarrow 0.3492$	$a_6 \rightarrow 0.0051$
$a_2 \rightarrow 0.0858$		$a_1 \rightarrow 0.1675$	$a_7 \rightarrow 0.0007$
			$a_8 \rightarrow 0.0018$

Since we are concerned with sample data at discrete interval of the independent variable, the convolution integral in eq.(4.15) is replaced by a summation for the short filter (Table 4.2) in the form :

$$T_k = \sum_{j=-2}^6 a_j \overline{\rho}_{k-j} \, ; k = 0, 1, \ldots \ldots 8. \qquad (4.16)$$

This signifies that a running weighted average of the input resistivity sampled data with the filter coefficients a_j gives transform value T_k at the sample point where digitized apparent resistivity is $\bar{\rho}_k$. Long filter (Table 4.3) may also be similarly used.

Now with the procedure laid down for the first step (Koefoed, 1968) of Koefoed's method, through filter theory we can easily compute $T(y)$ from the $\bar{\rho}(x)$ values. Ghosh's filter method (Ghosh, 1971a) together with Koefoed's modified approach (Koefoed, 1970) for the second step provides us with a fairly accurate combination for evaluation of layer parameters. This "Ghosh-Koefoed" method uses observed VES curve, sampled in a specific manner, to calculate resistivity transform through Ghosh's filter. Resistivity transform is then utilized to compute layer parameters through Pekeris (1940) recurrence relation.

The procedure consists of the following steps :

1. Apparent resistivity $\bar{\rho}$ vs. r known.

2. Calculate $T(m)$ through digitized sample value of $\bar{\rho}$ by convolving with Ghosh's filter coefficients:

$$T_k = \sum_{j=-2}^{6} a_j \bar{\rho}_{k-j} ; \quad \text{draw } T(m) \text{ curve.} \tag{4.17}$$

3. Assume an asymptotic approximation to the first part by a two-layer curve and accept these as parameters for top layer. This is based on the relation between resistivity transform and modified kernel function :

$$G_i = \frac{T_i - \rho_1}{T_i + \rho_1}$$

4. Reduce $T(m)$ to the lower boundary plane through (Pekeris, 1940):

$$T_{i+1} = \frac{T_i - \rho_i \tanh(mh_i)}{1 - T_i \tanh(mh_i)/\rho_i}$$

5. Continue operations (3) and (4) until the entire transform curve is exhausted and layer parameters obtained. Koefoed (1979) has given a complete automatic program for evaluation of layer parameters with digitized $\bar{\rho}$ values as the input.

A forward modelling algorithm based on Koefoed's (1979) eight sampling points per decade is computer coded in FORTRAN-77 on HP712/Silicon graphics Origin 200 system for systematic generation of synthetic VES curves. The FORTRAN source code of the program is given in Appendix 4.1. This can be used for generation of multi-layer

Schlumberger VES curves without any restriction, for use in the indirect interpretation of field curves.

As regards the direct approach to get the layer parameters, the inversion of resistivity field data is needed. The theoretical background for some useful resistivity inversion methods is prepared in the sections to follow.

4.5 One-Dimensional Resistivity Inversion by Ridge Regression

Many Geophysical models can be represented by the continuous functional relationship:

$$C = G\ (X,P) \tag{4.18}$$

.where,

    G = a function of model parameter P and known system parameter X,
    C = computed data from G.
For one-dimensional resistivity inversion data for layered earth model:
    P = layer resistivity and thickness,
    X = electrode separation,
    C = apparent resistivity.
In many cases $C = G\ (X, P)$ is linear i.e.:

$$C = AP \tag{4.19}$$

where,

    A = co-efficient matrix (Jacobian matrix)

If C can be the observed data, A is calculated from the forward problem and if A^{-1} exists then P(model parameters) can be calculated as:

$$P = A^{-1} C \tag{4.20}$$

For resistivity problems the system is quasi-linear. The non-linear sounding equation for apparent resistivity is linearized by Taylor Series expansion of $G\ (X, P)$ about the point (X, P^0) at each electrode separation. Retention of the first order terms only leads to a linear set of N equations in M unknowns given by:

$$\dot{} \Delta G = A\ \Delta P \tag{4.21}$$

where,

$$\Delta G_i = G\ (P, X_i) - G\ (P^0, X_i), \quad i = 1, 2, \ldots\ldots, N$$

and

$$[A]_{ij} = \frac{\partial G}{\partial P_j}(P, X)$$

where,

$$P = P^0$$

and

$$X = X_i$$

and

$$\Delta P_j = P_j - P_j^0 \quad j = 1, 2, 3, ..., M$$

Solution of the equation (4.21) can be obtained by least square method by multiplying the equation by A^T i.e.:

$$(A^T A)\Delta P = A^T \Delta G$$

The estimated model perturbation hence can be given as:

$$\Delta P = (A^T A)^{-1} A^T \Delta G \tag{4.22}$$

If $(A^T A)$ is singular, inversion does not exist. Damped least square technique is used to overcome this problem.

Eq. (4.22), therefore, gets modified into:

$$\Delta P = (A^T A + KI)^{-1} A^T \Delta G \tag{4.23}$$

where,

$K =$ damping factor,

$I =$ identity matrix.

In the ridge regression method the eigen values of the matrix $(A^T A + KI)$ are $\lambda + K$. The small eigen values are increased by a factor K which causes stability of the system. The larger eigen values are least affected because K is very small.

If K is large, ridge regression approaches gradient method and if K is small ridge regression method becomes a simple least square estimator equivalent to Newton-Raphson optimization technique.

4.6 Use of SVD Algebra in Solving the Ridge Regression Inversion Algorithm

A matrix 'A' can be written as a product of three other matrices U, V, Λ such that:

$$A = U\Lambda V^T$$

where,

$U_{n \times p} =$ data space eigen vector,

$V_{p \times p} =$ parameter space eigen vector,

$\Lambda = p \times p$ diagonal matrix containing atmost 'r' nonzero eigen values of 'A', with $r \leq p$.

These diagonal entries in Λ (λ_1, λ_2, λ_3, λ_4, ..., λ_p) are termed as Singular Values of 'A'.

The factorization is called Singular Value Decomposition (SVD) of 'A'. In SVD terms:

$$A^T A = U\Lambda V^T * V\Lambda U^T = V\Lambda^2 V^T \qquad (4.24)$$

And the equation (4.23) for Marquardt Levenberg ridge regression becomes:

$$\Delta P = V\Lambda_D^{-1} U^T \Delta G \qquad (4.25)$$

where,

$$\Lambda_D^{-1} = \frac{\Lambda}{(\Lambda + K)^2}$$

Thus, the matrix operation required for ridge regression technique has been substituted by SVD technique. This decreases the probability of singular matrix for which a Gauss-Jordan matrix inversion may fail. Thus, the algorithm becomes more stable. Quoting the remark for the stability of system it can be said that "this algorithm is technically sound and theoretically cannot fail".

4.7 Weighting and Scaling of Data for Ridge Regression

When the data are weighted, relative degree of importance is assigned to every value. Such weighting may be used to remove a bias inherent within the data or to bias the least squares fit so that it is more accurate in one area of the curve than the other.

If there is a large numerical difference between the values of data of the different regions of the curve, an undesirable bias may be introduced in the final solution. The bias is such as to cause the ridge regression estimator to be biased towards the larger values., which may be as accurate and may contain some important information. A ridge regression estimator and a least square estimator react to differences between the field curve and the curve generated from the estimated model. These differences would be greatest at the large array spacing merely because of the large numerical values of the curve in this region. The estimator might then give a good estimate of the resistivity of the lower half space but a poor estimate of the resistivity of the first layer. In general, it is desirable to weight each data point according to the noise in that data and, also, not give it a false degree of importance because of its large or small value in comparison with the other data points.

The weighting matrix M commonly used is $M = \sigma^2 N$. The matrix is a variance-covariance one of the data. A first order approximation assumes that the error in the data at one array spacing is unrelated to the error in the data in another spacing, so that the variance-covariance matrix becomes a diagonal matrix with elements σ_i^2. To determine σ_i^2 it is necessary to know the error in the data. Error in the data comes from various sources such as:

(a) Limited precision of instruction,
(b) Effect of lateral inhomogenities,
(c) Error in measuring spacing intervals.

The term σ^2 is called the problem variance. The procedure adopted to develop the program is that each data point has the same percentage standard deviation. It is further assumed that each point has a standard deviation 1% of its measured value. The problem standard deviation σ is then adjusted to the estimated noise level of the survey. Most resistivity surveys yield data that is accurate within 5% of its measured value.

To determine the weighting within estimator, each side of equation (4.21) is being multiplied by $N^{-\frac{1}{2}}$. Neglecting the error value, we can write:

$$N^{-\frac{1}{2}} \Delta G = N^{-\frac{1}{2}} A \Delta P \qquad (4.26)$$

The solution obtained by equation (4.26) is a weighted least-square solution. The weighted least square estimator is given by:

$$\Delta P = (A^T N^{-1} A)^{-1} A^T N^{-1} \Delta G \qquad (4.27)$$

The corresponding Ridge Regression estimator is given as:

$$\Delta P = (A^T N^{-1} A + KI)^{-1} A^T N^{-1} \Delta G \qquad (4.28)$$

Thus, eq.(4.23) of ridge regression estimator becomes eq. (4.28) after weighting.

SCALING: Before adding the factor K, it is convenient to scale the matrix $(A^T N^{-1} A)$ so that the diagonal elements have value of 1.0. The scaled matrix $(A^T N^{-1} A)^S$ and the scaled vector $(A^T N^{-1} \Delta G)^S$ are defined as:

$$(A^T N^{-1} A)_{ij}^S = \frac{(A^T N^{-1} A)_{ij}}{[(A^T N^{-1} A)_{ij}]^{\frac{1}{2}} [(A^T N^{-1} A)_{jj}]^{\frac{1}{2}}} \qquad (4.29)$$

$$(A^T N^{-1} \Delta G)_j^S = \frac{(A^T N^{-1} \Delta G)_j}{(A^T N^{-1} A)_{jj}} \qquad (4.30)$$

Defining diagonal scaling matrix with elements:

$$D_{ij} = 0, i \neq j$$

$$D_{ii} = [(A^T N^{-1} A)_{ii}]^{\frac{1}{2}}$$

and thus rewriting the equation (4.27) as:

$$\Delta P = D(DA^T N^{-1} AD)^{-1} DA^T N^{-1} \Delta G \qquad (4.31)$$

Ridge regression estimator becomes:

$$\Delta P = D(DA^T N^{-1} AD + KI)^{-1} DA^T N^{-1} \Delta G \qquad (4.32)$$

Eq. (4.32) is the estimator that gives the biased best fit to Schlumberger sounding data. The program written for *weighted ridge regression* uses eq.(4.32) to estimate ΔP.

A computer software is developed using weighted ridge regression technique for the inversion of Schlumberger VES data. The program is coded in FORTRAN 77 on HP712/ Silicon Graphics Origin 200 systems. The source code is given in Appendix 4.2.

4.8 One Dimensional Resistivity Inversion by Evolutionary Programming (EP)

Genetic algorithm (GA) has emerged as a powerful tool in the optimization theory in the late 1980s. This method has been used to optimize fitness function which arises in various ramification of science and engineering. It involves a learning process and hence classed, under the application part of artificial intelligence with the advent of parallel computing system, widely used to solve complex non-linear and non-local optimization problems.

GA (Goldberg and Richardson, 1987; Goldberg and Segrest, 1987; Goldgerg 1989; Ingber and Rosen, 1992; Nix and Vose, 1992; Sen and Stoffa, 1995) belongs to the group of random search methods such as Monte Carlo, Simulated Evolution, Simulated Annealing (SA) and Very Fast Simulated Annealing (VFSA) technique (Fogel et al., 1966; Rubinstein, 1981; Fogel, 1988; Schneider and Whitman, 1990; Sen and Stoffa, 1991; Stoffa and Sen, 1992; Chundru et al., 1995; Chundru et al., 1996). The utility of random search method lies in its ability to explore the model space extensively. In GA the search is random but is guided by stochastic process which helps the system to learn the minimum path leading to the solution.

However, one of the fundamental difficulties of GA was pointed out by Fogel (1988). The difficulty lies in the premature convergence as after successive generations, the entire population converges to a set of coding such that the crossover no longer generates any new chromosomes. This may happen even before finding an optimal solution. Although mutation allows for diversity, the mutation rate is usually low, so that practically no improvement can be achieved in the final generation of the population. This problem can be solved by Evolutionary Programming (EP). Based on the concept of EP, a program is developed in FORTRAN 77 on HP712/ Silicon Graphics Origin 200 systems. The source code of the algorithm is presented in Appendix 4.3.

4.8.1 Basic Theory of Evolutionary Programming as Applied to Resistivity Inversion

In D.C. resistivity sounding we are concerned with the parameters like $\rho_1, \rho_2,..., \rho_n$ and $h_1, h_2,..., h_{n-1}$. For each parameter we have a pair of bounds, i.e., the upper and lower limits. We intend to find the exact solution within the specified domains. EP does this by using three main steps, namely, generation of population, computation of fitness and mutation.

(a) Generation of population: In the first step n real coded individuals in the population are generated randomly within the specified bounds. Two important criteria for the generation of population are the population size and randomization seed number. The choice of population size and random seed number depends on the desired computational efficiency. The individuals of the population are randomly created by considering 'n' values between upper and lower limits of each parameter. The number of individuals generated can be shown in a matrix form given below:

$$I = \begin{bmatrix} I_{11} & I_{21} & \cdots & I_{n1} \\ I_{12} & I_{22} & \cdots & I_{n2} \\ \cdot \cdot & \cdots & \cdots & \cdots \\ I_{1p} & I_{2p} & \cdots & I_{np} \end{bmatrix} \qquad (4.33)$$

where suffix p represents parameter and suffix n represents number of values for each parameter.

(b) Computation of fitness: Fitness function of each newly generated individuals of the population is calculated by using the concept of chi-square error.

The chi-square error is defined as:

$$\text{chi - square error} = \frac{1}{N} \sum_N \frac{(\overline{\rho}_0 - \overline{\rho}_c)^2}{\overline{\rho}_c} \qquad (4.34)$$

where,

$\overline{\rho}_0$ = observed apparent resistivity

$\overline{\rho}_c$ = computed apparent resistivity

N = number of observation.

Here, $\overline{\rho}$ is given by the expression (Stefanescu and Schlumberger, 1930):

$$\overline{\rho} = r^2 \int_0^\infty T(m) J_1 (mr) m \, dm \qquad (4.35)$$

where,

r = half the distance between the current electrodes

$J_1(mr)$ = Bessel function of first order

m = integration variable

$T(m)$ = electrical impedance, defined in Koefoed (1970) as the resistivity transform function of layer resistivities and thicknesses.

Instead of using *rms* error, chi-square error is used to compare the results with those obtained by SVD algorithm.

(c) Mutation: In this step an equal number of individuals are generated by perturbing each member of the population by step function mutation. The mutation value is assigned depending on the fitness. After several tests, the probability of mutation P_m is found to yield better convergence if a step like distribution is used as given below:

$P_m = 0.999$ if $\overline{\varepsilon} \leq 0.5$

$= 0.985$ if $0.5 < \overline{\varepsilon} \leq 1.0$

$= 0.97$ if $1.0 < \overline{\varepsilon} \leq 2.0$

$= 0.95$ if $2.0 < \overline{\varepsilon} \leq 5.0$

$= 0.87$ if $5.0 \leq \overline{\varepsilon} \leq 10.0$

$= 0.77$ if $10.0 < \overline{\varepsilon} \leq 25.0$

$= 0.65$ if $25.0 < \overline{\varepsilon} \leq 50.0$

$= 0.5$ if $50.0 < \overline{\varepsilon} \leq 100.0$

$=$ random value generated by using same seed value used for population generation   if $\overline{\varepsilon} > 100.0$

where $\overline{\varepsilon}$ is the chi-square error.

In the next step of mutation, a random number is generated again by using same seed value. If the number is greater than 0.5, then mutated individuals are calculated by multiplying the individuals with probability P_m. Otherwise, it is found out by dividing the individuals with P_m.

The modified n models after a particular iteration are mixed with those from the previous iteration. $2n$ models are arranged in the decreasing order of fitness value. The best n models are retained for the next iteration. The process is repeated until the population converges to a high fitness value.

4.9 Comparative Analysis of Weighted Ridge Regression and Evolutionary Programming Techniques

Standardization of any new technique needs comparison with existing one. There are several calculus based reconstruction techniques available viz., Singular Value Decomposition (SVD), Ridge Regression, Weighted Ridge Regression, etc. (Marquardt, 1970; Inman et al., 1973; Inman, 1975; Bichara and Lakshmanan, 1976; Jupp and Vozoff, 1975; Johansen, 1977; Constable et al., 1987). All the schemes are linear inverse schemes. However, we have chosen the results obtained by SVD for the purpose of comparison. We classify our studies into two groups to avoid uncertainty in interpretation, namely, synthetic model studies and real field data analysis.

4.9.1 Synthetic Model Studies

Since numerical case studies give perfect solution we will first establish the efficacy of our program using a three-layer and five-layer earth models. The model parameters for both cases are shown in Table 4.4

Table 4.4

Synthetic Case	ρ_1	ρ_2	ρ_3	ρ_4	ρ_5	h_1	h_2	h_3	h_4	h_5
3-layer	10	1000	100	–	–	2	10	∞	–	–
5-layer	10	1000	100	500	50	1.5	7	9	20	∞

(i) Three-layer Case: The three-layer curve considered here is of H-type. The curve was then perturbed 10% and reconstructed by using both SVD and EP. For SVD we have given a fixed initial guess with values close to the actual parameters. In case of EP, a range is given for each parameter. Range for each parameter is as follows:

5 ohm $- m \le \rho_1 \le 15$ ohm $- m$; 995 ohm $- m \le \rho_2 \le 1005$ ohm $- m$; 95 ohm $- m \le \rho_3 \le 105$ ohm $- m$; $1m \le h_1 \le 3m$; $9m \le h_2 \le 11m$.

We have used a population size of 100 and randomization seed 2. The parameters obtained by using SVD and EP are shown in Table 4.5. Both the schemes give results quite close to the actual one. In Fig. 4.1 (a), a composite plot of apparent resistivity versus $AB/2$ for the original three-layer case along with that found out by SVD and EP are presented. Figure. 4.1(b) shows the chi-square error versus number of iteration plot for both SVD and EP. The chi-square error for EP converged at an iteration value equal to 4 which is much faster than SVD.

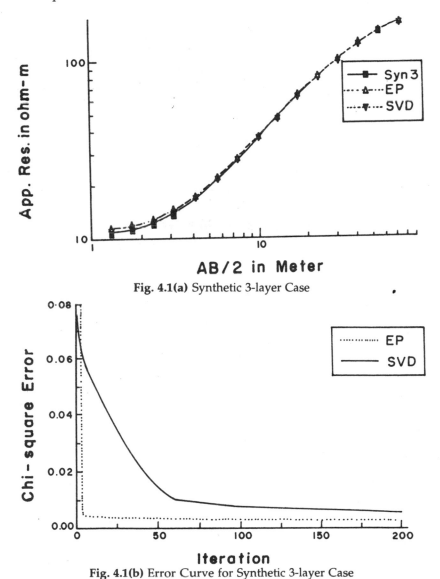

Fig. 4.1(a) Synthetic 3-layer Case

Fig. 4.1(b) Error Curve for Synthetic 3-layer Case

Table 4.5

Algorithm	ρ_1	ρ_2	ρ_3	h_1	h_2	h_3
SVD	10.49	942.07	129.37	2.07	9.6	∞
EP	10.63	977.99	105.71	2.11	9.87	∞

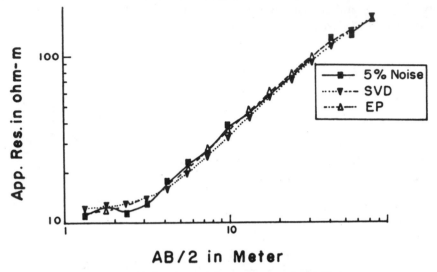

Fig. 4.2(a) Synthetic 3-layer Case with 5% Noise

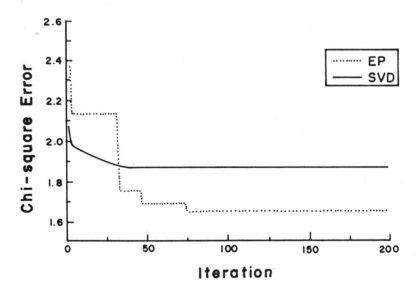

Fig. 4.2(b) Error Curve of Synthetic 3-layer Case with 5% Noise

Then, a ± 5% and ± 10% Gaussian noises are introduced to apparent resistivity data and reconstructed by using both SVD and EP. The results are presented in Table 4.6 and the curves plotted are shown in Figs. 4.2(a) and 4.3(a) with the corresponding error curves shown in Figs. 4.2(b) and 4.3(b). It is evident for both the cases that EP gives a better solution. Although both algorithms can handle a ± 5% and ±10% noisy data, EP gives better precision.

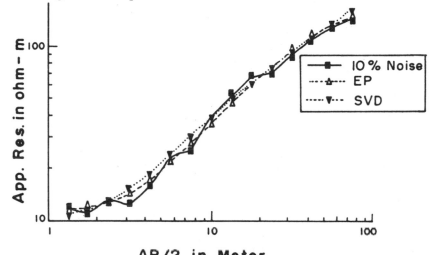

Fig. 4.3(a) Synthetic 3-layer Case with 10% Noise

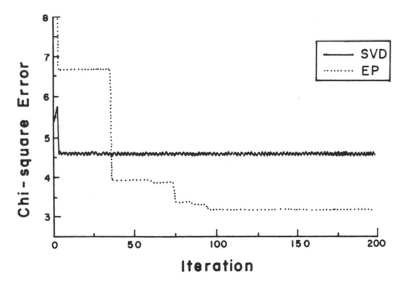

Fig. 4.3(b) Error Curve of Synthetic 3-layer Case with 10% Noise

Table 4.6

Algorithm	\multicolumn{6}{c}{3-layer Synthetic Case with ± 5% Noise}	\multicolumn{6}{c}{3-layer Synthetic Case with ± 10% Noise}
	ρ_1	ρ_2	ρ_3	h_1	h_2	h_3	ρ_1	ρ_2	ρ_3	h_1	h_2	h_3
SVD	11.5	2947.1	314.7	2.6	2.2	∞	10.1	629.6	205.7	1.6	30.1	∞
EP	10.7	1002.4	110.0	2.2	10.9	∞	10.9	900.0	90.0	2.2	9.0	∞

(ii) Five-layer Case: The five-layer case considered is of *HKH*-type. Here the same procedure was followed as in the previous case. The apparent resistivities were perturbed 10%. The educated guess for the SVD inversion was taken quite close to the actual values of the layer parameters. The range for each parameter for EP are as follows:

5ohm-m $\leq \rho_1 \leq$ 15 ohm-m; 995 ohm-m $\leq \rho_2 \leq$ 1005 ohm-m; 95 ohm-m $\leq \rho_3 \leq$ 105 ohm-m;

495 ohm-m $\leq \rho_4 \leq$ 505 ohm-m; 45 ohm-m $\leq \rho_5 \leq$ 55 ohm-m;

1m $\leq h_1 \leq$ 2m; 6m $\leq h_2 \leq$ 9m; 8m $\leq h_3 \leq$ 10m; 18m $\leq h_4 \leq$ 22m.

The population size and the seed for randomization considered are same as in the previous case. Table 4.7 gives the layer parameters obtained by using SVD and EP. A composite plot of apparent resistivity versus *AB*/2 for original five-layer case and that found out by SVD and EP are shown in Fig. 4.4(a). The reconstructed apparent resistivity obtained by using both SVD and EP are found to be matching with the apparent resistivity of the original plot. But from the chi-square error versus iteration

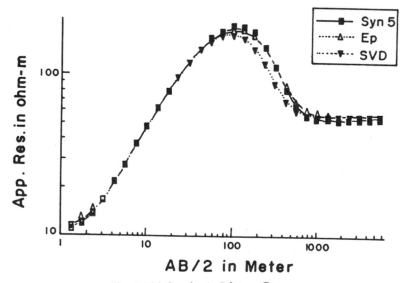

Fig. 4.4(a) Synthetic 5-layer Case

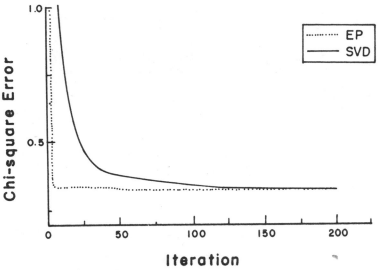

Fig. 4.4(b) Error Curve of Syn. 5-layer Case

plot shown in Fig. 4.4(b) it is observed that once again the EP converges to a value less than SVD. From the above two synthetic case studies it can be stated that both SVD and EP performs well although from the error trend it appears that the EP performs better than SVD. The studies carried out above also demonstrate the efficacy and robustness of our algorithm. This is further testified with the help of actual field study from Tentulia, Midnapur district, West Bengal (WB), INDIA outlined in the following section.

Table 4.7

Algorithm	ρ_1	ρ_2	ρ_3	ρ_4	ρ_5	h_1	h_2	h_3	h_4	h_5
SVD	10.64	1176.9	70.53	540.6	52.78	1.59	6.12	8.02	19.03	∞
EP	10.52	979.41	110.0	549.67	52.99	1.56	6.32	8.16	18.64	∞

4.9.2 Real Field Data Analysis

The site chosen for field studies is in the bank of river Kasai in Midnapur district, WB, INDIA. The main lithotypes of the area are newer alluvium, older alluvium and laterites. The sand thickness varies from 10 m to 15 m in the Tentulia area followed by a layer of clay. The aquifer is found to be an unconfined one. In the area 11, VES profiles were carried out of which three profiles are considered here to demonstrate the effectiveness of our algorithm. The location map of Tentulia is shown in Fig. 4.5(a). Figure. 4.5(b) shows the selected VES points.

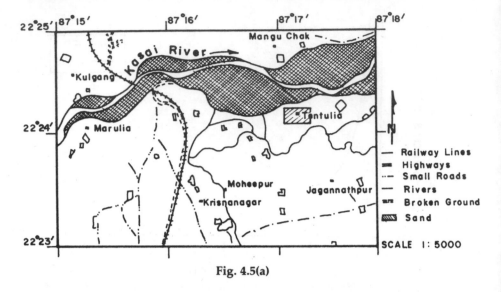

Fig. 4.5(a)

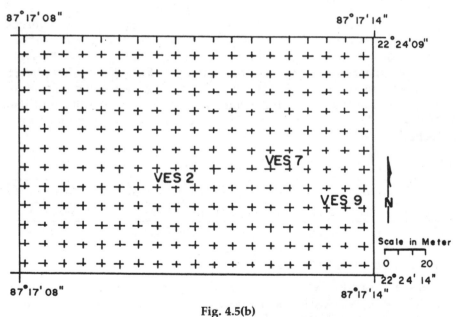

Fig. 4.5(b)

VES-2 was first interpreted by curve matching technique using auxiliary Ebert charts. The parameters obtained are as follows :

$\rho_1 = 110$ ohm-m; $\rho_2 = 520$ ohm-m; $\rho_3 = 20$ ohm-m; $h_1 = 1$ m; $h_2 = 6$ m; $h_3 = \infty$.

From the apparent resistivity versus $AB/2$ plot it is seen to be a K-type curve. The parameter values obtained by curve matching technique is used as initial guess for SVD program. For the EP program, we have given a range to all the parameters as in the synthetic cases. The parameters obtained by both the algorithms are presented in Table 4.8. In all the field cases, the population size and randomization seed number was considered as 100 and 2, respectively. The curves obtained by SVD and EP are shown in Fig. 4.6(a) and the chi-square error is shown in Fig. 4.6(b).

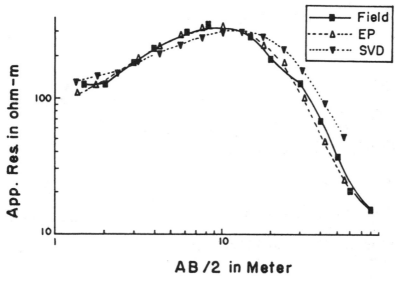

Fig. 4.6(a) VES - 2

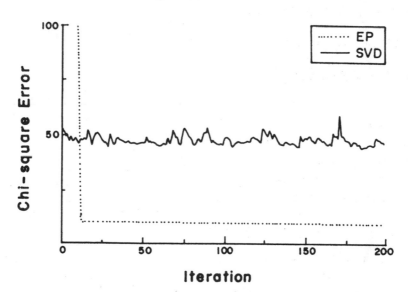

Fig. 4.6(b) Error Curve for VES - 2

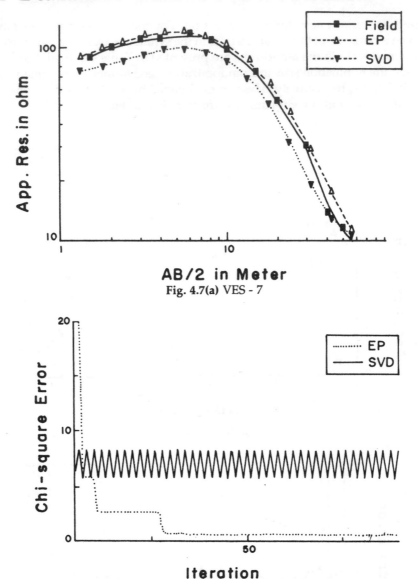

Fig. 4.7(a) VES - 7

Fig. 4.7(b) Error Curve for VES - 7

VES-7 is also a *K*-type curve. The parameters obtained by both the algorithms are shown in Table 4.8. The composite plot for apparent resistivity versus *AB*/2 is shown in Fig. 4.7(a) and the chi-square error versus number of iteration plot is shown in Fig. 4.7(b). The same procedure is followed for VES-9 which is of *HK*-type. In Table 4.8, the parameters obtained by using both the techniques are shown. The composite plot of apparent resistivity versus *AB*/2 and chi-square error versus iteration is shown in Figs. 4.8(a) and (b).

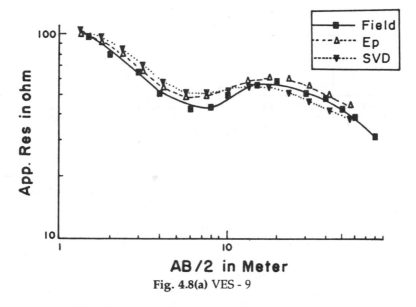

Fig. 4.8(a) VES - 9

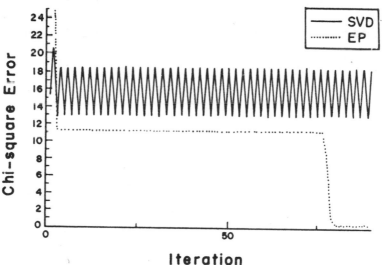

Fig. 4.8(b) Error Curve for VES - 9

Analysis of synthetic and field studies suggest that both SVD and EP are effective in interpreting apparent resistivity versus $AB/2$ curves. However, from the composite plots of apparent resistivity versus $AB/2$ and chi-square error versus number of iteration for all the cases it is undoubtedly established that results obtained by using EP are better than those estimated by SVD.

Table 4.8

VES	Algorithm	ρ_1	ρ_2	ρ_3	ρ_4	h_1	h_2	h_3	h_4
2	SVD	119.63	523.16	21.16	–	1.31	6.04	∞	–
	EP	77.78	2953.66	13.33	–	0.84	0.90	∞	–
7	SVD	66.83	135.34	75.12	6.96	0.88	2.02	4.12	∞
	EP	69.59	141.08	46.55	4.88	0.52	3.61	7.98	∞
9	SVD	105.88	16.15	80.0	29.4	1.2	0.94	5.0	∞
	EP	106.26	28.77	617.92	32.49	0.99	2.85	0.76	∞

These algorithms are routinely used for Schlumberger VES data interpretation for the exploration of ground water. As a matter of fact, curve matching technique elaborated in section 3.7 does the initial parameterization in geosounding data interpretation to obtain the final layer parameters. Chapter 5 deals with several field examples wherein evaluation of aquifer parameters is done through indirect (curve matching and use of forward algorithm given in Appendix 4.1) and direct (resistivity inversion through algorithms given in Appendices 4.2 and 4.3) approaches. In case, one has to choose between SVD and EP, EP is preferred for the final evaluation of layer resistivities and thicknesses.

5 Application in Ground Water

5.1 Introduction

There are many applications of Schlumberger Vertical Electrical Sounding (VES) in ground water prospecting leading to evaluation of subsurface water resource of an area. Prospecting, development and management of the resources will require systematic studies on the following points.

(1) Utilization of remote sensing data and existing geological information to prepare a surface geological map followed by water-table map of the area.

(2) VES data collection, interpretation (curve matching and resistivity inversion), preparation of geoelectric section, correlation with available lithology (if any) and recommendation of the drilling points based on inferred geological section.

(3) Electrical logging of the boreholes immediately after drilling to delineate the porous and permeable water-bearing zones for a judicious strainer positioning, for an ideal yield.

(4) Pumping test, sand analysis and determination of aquifer parameters (transmissibility and storativity) for computation of yield leading to a planned utilization of the available groundwater resources reserve.

(5) Groundwater quality studies for suitability of the water for drinking, industrial or agricultural purposes as the case may be.

Schlumberger VES, detailed in the present book, is meant for carrying out the second step of operation mentioned above. The procedure consists of the collection of sets of apparent resistivity ($\bar{\rho}$ in ohm-m) data versus half-electrode separation ($AB/2$ in metres) as per plans given in the typical Schlumberger layout (Table 2.1). The data are plotted on double-log transparent graphsheet of modulus 62.5 mm and the field curve obtained. The layer parameters (thicknesses and resistivities) are calculated through curve matching technique (Sec. 3.7) using two-layer master curves and Ebert charts (Figs. 3.1, 3.8 and 3.9). These approximate values of layer parameters are used to generate the theoretical curve for comparison with the field curve (Forward algorithm, Appendix 4.1). The parameters after the final match are taken as results of preliminary interpretation. Further refinement is achieved through treatment of above data as initial guess for the resistivity inversion algorithms (Appendices 4.2 and 4.3).

The application of Schlumberger VES in ground water under different geological conditions are briefly outlined in the following sections.

5.2 A Costal Area Problem

5.2.1 Geology of the Area

Figures 5.1 and 5.2 show the application of Schlumberger VES around Jaldha-Digha Coastline situated on the shore of the Bay of Bengal (Fig. 5.6) located at a distance of about 150 km west of Calcutta. The area is bounded by 87°30′ and 87°25′E and by 21°35′ and 21°40′N.

The area under investigation forms a coast of emergence. The ground surface is more or less flat (except for a few sand dunes) with a gentle slope towards the sea. Here, the shoreline is of the "sandy type" and has been formed against the unconsolidated sediments of Upper Tertiary age which dip at low angles to the south-southwest. The area is covered by unconsolidated clays, silt and sand deposited in Recent times by marine transgression. The subsurface formations consist of alternate sand and clay layers. Although no saline invasion is reported from the area, existence of saline water-bearing pockets is confirmed from existing borehole logs.

5.2.2 Results of VES

Several Schlumberger VES were carried out around Jaldha coast and the data were interpreted by curve matching and verified through forward approach. The interpreted layer parameters were correlated with available borehole data and a probable geological section prepared (Patra, 1967). A low resistivity 0.7 ohm-m (corresponding to a saline water-bearing fine sand as shown in the legend, Fig. 5.1), 19m thick layer at a depth of about 14 m from the surface has been located at the point for VES-1(Fig. 5.1). This, however, does not extend laterally inland even upto VES-2 and therefore, act as an isolated pocket of sand. The probable lateral extension of the pocket has been shown qualitatively on the diagram.

Figure. 5.2A gives the VES data from Digha coast plotted on a double-logarithm graphsheet of modulus 62.5 mm and denoted as VES-1 to VES-6. The values of interpreted layer parameters correlated with available borehole log are used to infer the lithology. The geological section prepared from VES and borehole data (Fig. 5.2B) is reproduced from Patra and Bhattacharya (1966).

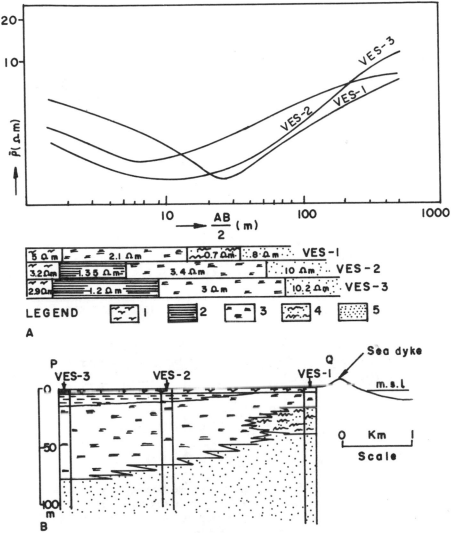

Fig. 5.1 Schlumberger VES curves representing section PQ.
B.   Geological section along PQ prepared from the information given
in A. Key: 1 = alluvium (3 - 5 Ω m);
2 – plastic clay (1 - 2 Ω m); 3 = silt and sand (3 - 4 Ω m);
4 = salt water bearing fine sand (< 1 Ω m);
5 = medium clay sand (8 - 12 Ω m)

5.3   A Soft Rock Area Problem

5.3.1   Geology of the Area

The area around Tata Metalik sponge iron works (shaded) was surveyed

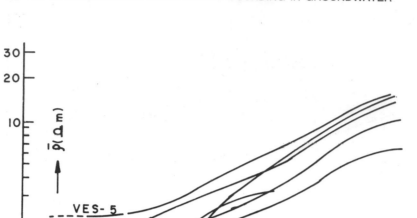

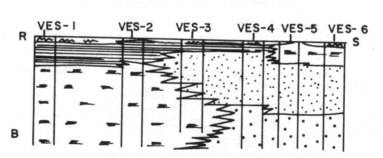

Fig. 5.2 Schlumberger VES curves representing section RS. Key same as in Resistivity more than 15 Ω m for medium to coarse sand.
B.   Geological section along RS prepared from the information obtained with the curves presented in Fig. 5.2A.

in detail using Schlumberger VES (Fig. 5.3). Here, the low-lying areas, forming a part of Kasai river valley are surfaced by Newer Alluvium of Pleistocene to Recent in age. This formation with a maximum thickness of 125 m is a series of floodplain deposits in valleys cut into the Older Alluvium. The Newer Alluvium is a medium to fine grained deposit, mainly of silt and silty clays, but sometimes with good clean sands in the palaeochannel areas (Fig. 5.3). The tubewells drilled within Newer Alluvium have normally been observed to have a gradual reduction in yield with time.

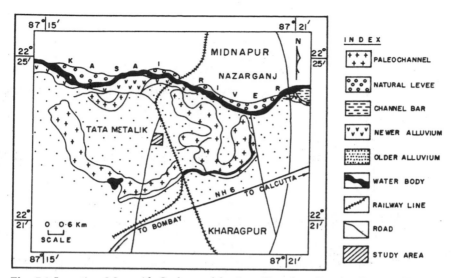

Fig. 5.3 Location Map with Geology of the Area Based on Remote Sensing Data.

5.3.2 Results of VES

A total of 30 VES curves were obtained from the area referred to as VES-

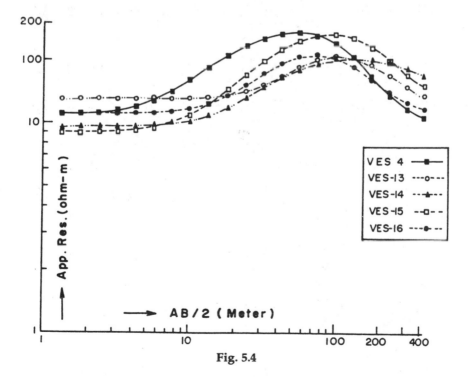

Fig. 5.4

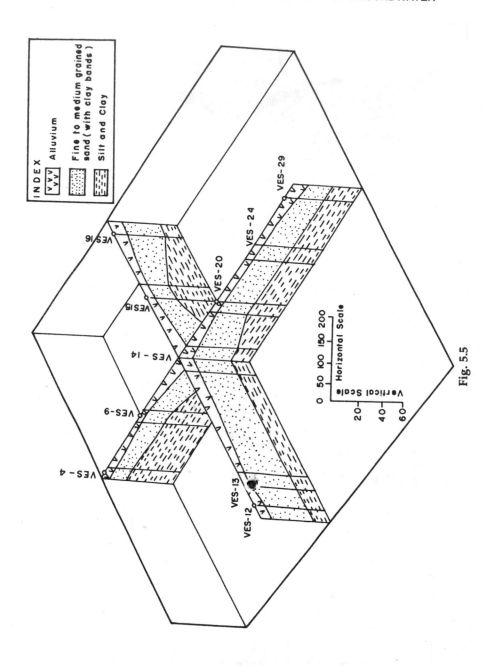

Fig. 5.5

1, VES-2........VES-30. While VES-1, VES-3 and VES-5 represent four-layer *AK*-type case, the rest 27 curves represent three-layer *K*-type cases. Some representative VES curves (VES-4, VES-13, VES-14, VES-15 and VES-16) are reproduced (Fig. 5.4) here from Patra (1997).

The apparent resistivity data are interpreted by curve matching technique followed by resistivity inversion to get the final layer parameters (thicknesses and resistivities). The resistivity values of the various layers at all the thirty points are used to infer the subsurface lithology after correlation with available borehole data from the area. Finally, the interpreted geoelectric section is utilized to prepare a subsurface geological section presented in Fig. 5.5.

Thus, starting from the VES data, the geological section is prepared which shows that a thick fine to medium grained sand exists below a thin alluvial cover and rests over a thick impervious clay (Fig. 5.5).

5.4 A Hard Rock Area Problem

5.4.1 Geology of the Area

The area of study around Noamundi iron ore mines is shown in Fig. 5.6 with latitudes and longitudes given on the diagram. Geologically, the

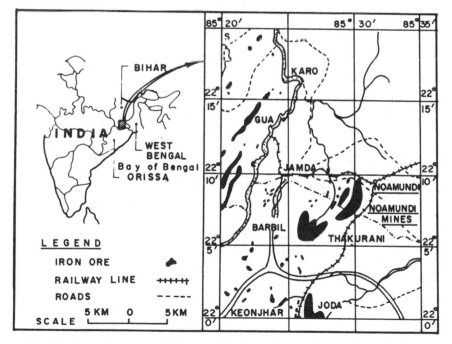

Fig. 5.6 Location map of Noamundi mines. Area under study is shaded.

formations belong to Archeans comprising a metamorphosed sedimentary rock system, in India, developed in the eastern states of Bihar and Orissa (Fig. 5.6) with a thick group of ferruginous sediments. It consists essentially of iron-bearing sediments-phyllites, tuffs, lavas, quartzites and limestones designated as iron ore series—resting unconformably on an older metamorphic series. In its petrogenesis, the series is believed to be akin to Lake Superior Precambrian iron-bearing formation of U.S.A.

The study area is located (Fig. 5.6) within the precambrian metamorphic sediments around Noamundi (85°30', 22°10') where rocks have suffered deformation in successive stages. The yield of ground water in these rocks depends on the presence of weathered pockets, joints and fractures varying widely within short distances, controlled largely by weathering. Schlumberger VES were conducted at seven points (P-1 to P-7) around the palletizing plant (Fig. 5.7). The curves at these points are represented by VES-1 to VES-7 (Figs. 5.8 and 5.9).

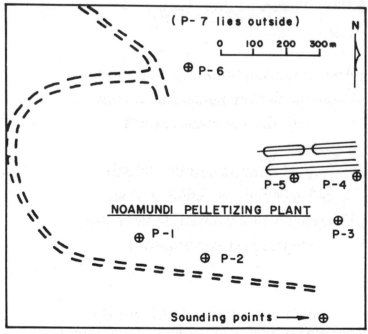

Fig. 5.7   Location of VES points around Noamundi pelletizing plant.

5.4.2   Results of VES

The seven soundings from the palletizing plant area (VES-1 to VES-7) are given in Figs. 5.8 and 5.9. The data interpreted through curve matching, forward algorithm and resistivity inversion (SVD and EP) give layer parameters, the values of which are given in-set the diagrams.

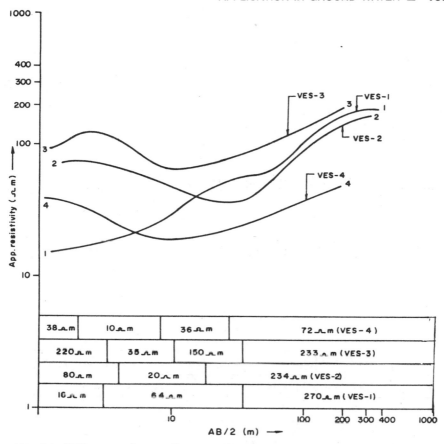

Fig. 5.8 VES curves from pelletizing plant area, Noamundi Iron Ore Mines. Geoelectric sections corresponding to VES curves are shown at the bottom.

Analysis of the results from the area (Patra, 1993) given in Figs. 5.8 and 5.9 indicates the presence of moderate resistivity (230-270 ohm-m) zones for VES-1, VES-2 and VES-3 within a depth of 20-40 m showing the probable presence of partly fractured saturated zone. VES-4, VES-5 and VES-6 show low resistivity values (35-64 ohm-m) at depths around 10-30 m showing a likely saturated zone. Low resistivity (35-64 ohm-m) at shallow depths (3-8 m) are also prevalent at P-1, P-3 and P-4.

Patra (1993) records that partly fractured saturated zones exist at almost all the points but the points P-3, P-4 and P-5 appear favourable from ground water potential point of view. A test borehole was recommended around these points upto a depth of about 100 m.

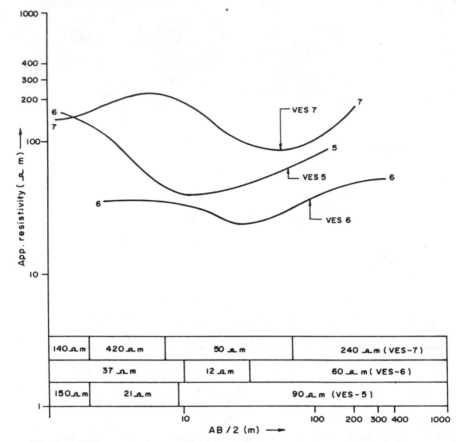

Fig. 5.9 VES curves from pelletizing plant area, Noamundi Iron Ore Mines. Geoelectric sections corresponding to VES curves are shown at the bottom.

References

Backus, G.E. and Gilbert, F. 1967. Numerical applications of a formalism for geophysical inverse problems. *Geophys. J.R. Astr. Soc.*, vol. 13, pp. 247-276.

Backus, G.E. and Gilbert, F. 1968. The resolving power of gross earth data. *Geophys. J.R. Astr. Soc.*, vol. 16, pp. 169-205.

Backus, G.E. and Gilbert, F. 1970. Uniqueness in the inversion of inaccurate gross earth data. *Phil. Trans. R. Soc. Lond. A.*, vol. 226, pp. 123-192.

Bhattacharya, P.K. and Patra, H.P. 1968. *Direct Current Geoelectric Sounding: Principles and Interpretation*. Elsevier, Amsterdam, 135 pp.

Bichara, M. and Lakshmanan, J. 1976. Fast automatic processing of resistivity sounding. *Geophys. Prospect.*, vol. 24, pp. 354-370.

Chundru, R.K., Sen, M.K., Stoffa, P.L. and Nagendra, R. 1995. Nonlinear inversion of resistivity profiling data for some regular geometrical bodies. *Geophys. Prospect.*, vol. 43, pp. 979-1004.

Chundru, R.K., Sen, M.K. and Stoffa, P.L. 1996. 2-D resistivity inversion using spline parameterization and simulated annealing. *Geophysics*, vol. 61, pp. 151-161.

Compagnie Generale De Geophysique. 1955. Abaque de sondage electrique. *Geophys. Prospect.*, vol. 3, (Suppl.3) 50 pp.

Compagnie Generale De Geophsique. 1963. *Master Curves for Electrical Sounding*. 2nd Ed. European Assoc. Exploration Geophysicists, *The Hague*, 49 pp.

Constable, S.E., Parker, R.L. and Constable, C.G. 1987. Occam's inversion: A practical algorithm for generating smooth models for electromagnetic sounding data. *Geophysics*, vol. 52, pp. 289-300.

Deppermann, K. 1973. An interpretation system for geoelectrical sounding graphs. *Geophys. Prospect.*, vol. 21, pp. 424-463.

Flathe, H. 1955. A practical method of calculating geoelectrical model graphs for horizontally stratified media. *Geophys. Prospect.*, vol. 3, pp. 268-294.

Fogel, D.B. 1988. An evolutionary approach to travelling salesman problem. *Biological cybernetics*, vol. 60, pp. 139-144.

Fogel, L.J., Owens, A.J. and Walsh, M.J. 1966. *Artificial Intelligence through Simulated Evolution*. John Wiley and Sons, New York.

Ghosh, D.P. 1970. *The Application of Linear Filter Theory to the Direct Interpretation of Geoelectrical Resistivity Measurements*. Doctoral Thesis, Tech. Univ. Delft, 125 pp.

Ghosh, D.P. 1971a. The application of linear filter theory to the direct interpretation of geoelectrical resistivity sounding measurements. *Geophys. Prospect.*, vol. 19, pp. 192-217.

Ghosh, D.P. 1971b. Inverse filter coefficients for the computation of apparent resistivity standard curves for a horizontally stratified earth. *Geophys. Prospect.*, vol. 19, pp. 769-775.

Goldberg, D.E. 1989. *Genetic Algorithms in Search, Optimisation and Machine Learning*, Addison-Wesley, Reading.

Goldberg, D.E. and Richardson, J. 1987. Genetic algorithms with sharing for multimodal function optimisation. *Proceeding of the Second International Conference on Genetic Algorithms*, J. J. Grefenstette(Editor), pp. 41-49.

Goldberg, D.E. and Segrest, P. 1987. Finite Markov chain analysis of genetic algorithms. *Proceedings of the Second International Conference on Genetic Algorithms*, J. J. Grefenstette(Editor), pp. 1-8.

Hoverstein, G.M., Dey, A. and Morrison, H.F. 1982. Comparison of five least square inversion techniques in resistivity sounding. *Geophys. Prospect.*, vol. 30, pp. 688-715.

Ingber, L. and Rosen, B. 1992. Genetic algorithms and simulated reannealing: A comparison. *Mathl. Comput. Modeling*, vol. 16, pp. 87-100.

Inman, J.R. 1975. Resistivity inversion with ridge regression. *Geophysics*, vol. 40, pp. 798-817.

Inman, J.R., Ryu, J. and Ward, S.H. 1973. Resistivity inversion. *Geophysics*, vol. 38, pp. 1088-1108.

Johansen, H.K. 1975. An interactive computer/graphic display-terminal system for interpretation of resistivity soundings. *Geophys. Prospect.*, vol. 23, pp. 449-458.

Johansen, H.K. 1977. A man/computer interpretation system for resistivity soundings over a horizontally stratified earth. *Geophys. Prospect.*, vol. 25, pp. 667-691.

Jupp, D.L.B. and Vozoff, K. 1975. Stable iterative methods for the inversion of geophysical data. *Geophysics*, vol 42, pp. 957-976.

Kalenov, E.N. 1957. *Interpretatsia Krivikh Verlikalnogo Electrichekogo Zondirovanya*. Gostoptekhizdat, Moskva, 470 pp.

Koefoed, O. 1965. A semidirect method of interpreting resistivity observations, *Geophys. Prospect.*, vol. 13, pp. 259-282.

Koefoed, O. 1968. *The Application of the Kernel Function in Interpreting Geoelectrical Resistivity Measurements*. Borntraeger, Berlin, 111 pp.

Koefoed, O. 1970. A fast method for determining the layer distribution from the raised kernel function in geoelectrical sounding. *Geophys. Prospect.*, vol. 18, pp. 564-570.

Koefoed, O. 1979. *Geosounding Principles, 1 - Resistivity Sounding Measurements*. Elsevier Scientific Publishing Co. Amsterdam, 276 pp.

Kunetz, G. 1966. *Principle of Direct Current Resistivity Prospecting*. Borntrager, Berlin, 106 pp.

Langer, R.E. 1933. An inverse problem in differential equations. *Am. Math. Soc. Bull.*, vol. 39, pp. 814-820.

Maillet, R. 1947. The fundamental equations of electrical prospecting. *Geophysics*, vol. 12, pp. 529-556.

Marquardt, D.W. 1970. Generalised inverse, ridge regression, biased linear estimation and non-linear estimation. *Technometrics*, vol. 12, pp. 591-612.

Meinardus, H.A. 1970. Numerical interpretation of resistivity soundings over horizontal beds. *Geophys. Prospect.*, vol. 18, pp. 415-433.

Mooney, H.M. and Wetzel, W.W. 1956. *The potential about a Point Electrode and Apparent Resistivity Curves for a Two-, Three - and Four-Layer Earth*. Univ. Minnesota, Mineapolis, 146 pp.

Mooney, H.M., Orellana, E., Pickett, H. and Tornheim, L. 1966. A resistivity computation method for layered earth models. *Geophysics*, vol. 31, pp. 192-203.

Nix, A.E. and Vose, M.D. 1992. Modelling genetic algorithms with Markov chains. *Ann. of Mathematics and Artificial Intelligence*, vol. 5, pp. 79-88.

Orellana, E. and Mooney, H.M. 1966. *Master Tables and Curves for Vertical Electrical Sounding over Layered Structures*. Intersciencia, Madrid, 193 pp.

Patra, H.P. 1967. On the possibility of saline water invasion around the Jaldha coast, West Bengal (India). *Geoexploration*, vol. 5, pp. 95-101.

Patra, H.P. 1993. Ground water potential within hard rocks around Noamundi iron mines, Bihar, India. *Hydrogeology of Hard Rocks*; Editors: Sheila and David Banks, Memoirs of the XXIVth Congress, IAH, ÅS (Oslo), Norway, pp 531-537.

Patra, H.P. 1997. *Report on Electrical Resistivity Survey and Pump Test for Ground Water in and around Tata Metaliks Ltd., Kharagpur*, Unpublished Report, 45 pp.

Patra, H.P. and Bhattacharya, P.K. 1966. Geophysical exploration for ground water around Digha in the coastal region of West Bengal, India. *Geoexploration*, vol. 4, pp. 209-218.

Pekeris, C.L. 1940. Direct method of interpretation in resistivity prospecting. *Geophysics*, vol. 5, pp. 31-46.

Rubinstein, R.Y. 1981. *Simulation and the Monte Carlo Method*. John Wiley and Sons, New York.

Schneider, W.A. and Whitman, W.W. 1990. Dipmeter analysis by a Monte Carlo technique. *Geophysics*, vol. 55, pp. 320-326.

Sen, M.K. and Stoffa, P.L. 1991. Non-linear one-dimensional seismic waveform inversion using simulated annealing. *Geophysics*, vol. 56, pp. 1624-1638.

Sen, M.K. and Stoffa, P.L. 1995. *Global Optimisation Methods in Geophysical Inversion*. Elsevier , Amsterdam, 281 pp.

Slichter, L.B. 1933. The interpretation of resistivity prospecting method for horizontal structures. *Physics*, vol. 8, pp. 307-322.

Stefanescu, S.S. and Schlumberger, C. and M. 1930. Sur la distribution electrique potentielle antour d'une prise de terre ponctnelle dans un terrain a cauches horizontales, homogenes et isotropes. *J. Phys. Radium*, vol. 7, pp. 132-141.

Stoffa, P.L. and Sen, M.K. 1992. Seismic waveform inversion using global optimisation. *Seis. Exp.*, vol. 1, pp 9-27.

Van Dam, J.C. 1965. A simple method for calculation of standard-graphs to be used in geoelectrical prospecting. *Geophys. Prospect*, vol. 13, pp. 37-64.

Van Dam, J.C. 1967. Mathematical denotation of standard graphs for resistivity prospecting in view of their calculation by means of a digital computer. *Geophys. Prospect.*, vol. 15, pp. 57-70.

Vozoff, K. 1958. Numerical resistivity analysis: horizontal layers. *Geophysics*, vol. 23, pp. 536-556.

Zohdy, A.A.R. 1965. The auxiliary point method of electrical sounding interpretation and its relationship to the Dar Zarrouk parameters. *Geophysics*, vol. 30, pp. 644-660.

Zohdy, A.A.R. 1973. *A Computer Program for the Automatic Interpretation of Schlumberger Sounding Curves over Horizontally Stratified Media*. U.S.G.S. Report, no. GD-74-017, NTIS, Springfield.

Zohdy, A.A.R. 1974. *Use of Dar Zarrouk Curves in the Interpretation of Vertical Electrical Sounding Data*. U.S.G.S. Bulletin. 1313-D, Washington, 41 pp.

APPENDICES

Appendix

Guidelines for Operating the Forward Algorithm Given in Appendix 4.1

Here, layer parameters are given as INPUT and apparent resistivity $(\bar{\rho})$ versus half electrode spacing $(AB/2)$ are obtained as OUTPUT.

Layer parameters are :
  (i) Number of layers
  (ii) Resistivities of layers
  (iii) Thicknesses of layers.

Feed the layer parameters in the following sequence:
  (i) Top layer parameters
  (ii) Then, going down step by step.

For the OUTPUT the following details are needed :
  (i) Number of samples of the output (maximum 35)
  (ii) First abscissa value = first point of the output.

For sampling at eight points per decade, the sample points (upto 35) are:
1.33, 1.77, 2.37, 3.16, 4.22, 5.62, 7.50,10.0
13.3, 17.7, 23.7, 31.6, 42.2, 56.2, 75.0, 100.0
... ...
13335.0, 17782.0, 23714.0

The first abscissa value : 1.33

SAMPLE INPUT:
  Number of layers = 3
  Layer Resistivities = 1.6, 45.0, 16.0
  Layer Thicknesses = 0.4, 4.5
  Number of Data Points = 12
  First Abscissa Value = 1.33

SAMPLE OUTPUT:

AB/2	APPARENT RESISTIVITIES
1.33	4.88

1.77	6.26
2.37	7.98
3.16	10.04
4.21	12.43
5.62	15.04
7.49	17.66
10.00	19.95
13.33	21.50
17.78	22.01
23.71	21.45
31.62	20.20
42.16	18.79
56.23	17.67
74.98	16.97
100.00	16.59
133.35	16.41
177.82	16.33
237.13	16.32
316.22	16.35

Appendix 4.1

```
c      Forward Program using Koefoed's method


c      Declaration of global variables
       real xval(50),rapf(50),elecsep(50)
       integer no_layer,no_param,i,j,k,nopnew,nopoint
       real ini_L(50)
       real x1,tt(50)


c      Open output file
       open(unit=9,file='calculat.dat',status='unknown')


c      User defined layer parameters
       write(*,*)'Enter no. of Layers :'
       read(*,*)no_layer
       no_param=2*no_layer-1
       write(*,*)'Enter Model Params:'
       do i=1,no_param
       read(*,*)ini_L(i)
       end do


c      Online User Input : number of data points and first abscissa value
       write(*,*)'enter no. of data points'
       read(*,*) nopoint
       write(*,*)'enter first abscissa'
       read(*,*)x1


c      Call subroutine 'forward' for calculating App.Res. vs. AB/2 data
       call forward(no_layer,nopoint,ini_L,bak_tt,x1,1)

       close(9)
       stop
       end


c      Subroutine 'forward' to generate App.Res. vs. AB/2 data from layer
       parameters subroutine forward(i9,n,bb,ss,x1,in)

       real bb(50), r(50), d(50), t(35), ss(50), f, y, x1
```

```
       real u, b, a1, a2, axis(50)
       integer i9,i8,n,m,in

       do i=1,i9
       r(i)=bb(i)
       d(i)=bb(i+i9)
       end do
       f=exp(alog(10.0)/8)
       i8=i9-1
       y=x1/822.8
       do j=1,34
       b=r(i9)
       do k=1,i8
       i=i9-k
       u=d(i)/y
       if(5.0-u)14,14,15
14     b=r(i)
       goto 13
15     a1=exp(u)
       a2=(a1-1.0/a1)/(a1+1.0/a1)
       b=(b+a2*r(i))/(1.0+a2*b/r(i))
13     end do
       t(j)=b
       y=y*f
       end do
       do ks=1,50
       ss(ks)=0.0
       end do
       do m=1,n
       b=r(i9)
       do k=1,i8
       i=i9-k
       u=d(i)/y
       if(5.0-u)46,46,45
46     b=r(i)
       goto 44
45     a1=exp(u)
       a2=(a1-1.0/a1)/(a1+1.0/a1)
       b=(b+a2*r(i))/(1.0+a2*b/r(i))
44     end do
       t(35)=b
       y=y*f
       s=0.0
       s=42.*t(1)-103.*t(3)+144.*t(5)-211.*t(7)+330.*t(9)-574.*t(11)
```

```
s=s+1184.*t(13)-3162.*t(15)+10219.*t(17)-24514.*t(19)+18192.*t(21)
s=s+6486.*t(23)+1739.*t(25)+79.*t(27)+200.*t(29)-106.*t(31)
s=s+93.*t(33)-38.*t(35)
s=s/10000.0
ss(m)=s
do j=1,34
t(j)=t(j+1)
end do
end do
if(n.lt.8)then
do m=1,n
axis(m)=exp((m*log(10.))/8
end do
else
do m=1,8
axis(m)=exp((m*log(10.))/8
end do
if(n.lt.16)then
do m=9,n
axis(m)=axis(m-8)*10
end do
else
do m=9,16
axis(m)=axis(m-8)*10
end do
if(n.lt.24)then
do m=17,n
axis(m)=axis(m-16)*100
end do
else
do m=17,24
axis(m)=axis(m-16)*100
end do
if(n.lt.32)then
do m=25,n
axis(m)=axis(m-24)*1000
end do
else
do m=25,32
axis(m)=axis(m-24)*1000
end do
if(n.gt.32)then
do m=33,n
```

```
axis(m)=axis(m-32)*10000
end do
end if
end if
end if
end if
end if
if(in.eq.1)then
do m=1,n
write(9,*)axis(m);ss(m)
write(*,*)axis(m),ss(m)
end do
end if

return
end
```

Guidelines for Operating the Inversion Program Using Weighted Ridge Regresion Given in Appendix 4.2

Here, initial guess of layer parameters and apparent resistivity ($\bar{\rho}$) versus half electrode spacing ($AB/2$) data are given as INPUT and final layer parameters and corresponding chi-square error are obtained as OUTPUT.

INPUT data required :

Layer parameter data :
    (i) Number of layers
    (ii) Initial guess of layer resistivities
    (iii) Initial guess of layer thicknesses

Feed the layer parameters in the following sequence :
    (i) Top layer parameters
    (ii) Then, going down step by step.

Other data :
    (i) Name of INPUT field acquired data file ($\bar{\rho}$ vs. $AB/2$)
    (ii) Number of data points in the INPUT file
    (iii) Name of OUTPUT file
    (iv) Number of iterations

OUTPUT:

The result is stored in the output file and the error after each iteration is stored in a file named 'error.out'.

SAMPLE INPUT:

Number of layers = 3
Name of INPUT file = ves12.dat*
Number of data points = 22
Name of OUTPUT file = svdout.dat
Initial choice of layer resistivities : 200.0
                                             30.0
                                           50.0
Initial Choice for layer thicknesses = 2.5
                                              11.5
Number of Iteration = 1000

[* Sample field data stored in File "ves12.dat" : (H - Type curve)

AB/2	$\bar{\rho}$
1.5	22.0
2.0	19.8
3.0	16.0
4.0	14.6
5.0	14.0
6.0	14.3
7.0	15.0
8.0	16.2
9.0	17.1
10.0	18.5
15.0	26.0
20.0	32.4
30.0	43.0
40.0	47.2
50.0	46.8
60.0	42.5
70.0	39.0
80.0	36.2
90.0	33.4
100.0	31.0
120.0	26.0
140.0	22.0]

SAMPLE OUTPUT:

Resistivity of Layer - 1 = 162.14
Resistivity of Layer - 2 = 9.03
Resistivity of Layer - 3 = 34.23

Thickness of Layer - 1 = 0.53
Thickness of Layer - 2 = 7.31

Chi-square Error = 0.19

Appendix 4.2

c   **Inversion of Apparent resistivity data by Weighted Ridge Regression aided by SVD**

```
c    Declaration of global variables
     real r(50),h(50),b(1200),x1,sum
     real cor(1200,1200),g(1200)
     real raf(1200),chi(1200),bini(1200),rat(1200)
     real d(1200,1200),tt(1200)
     real bt(1200,1200),c(1200,1200),kk
     real keep1(1200,1200)
     real rapf(40),xval(40),keep2(1200,1200)
     real bak(1200,1200),a(1200,1200)
     real e(1200),elecsep(40)
     real df(1200),prob(1200)
     real up(1200,1200),vp(1200,1200),qp(1200)
     real ut(1200,1200),qi(1200)
     real vqi(1200,1200),utd(1200),mmt(1200),sol(1200,1200)
     character*20 infile,outfile
     integer no,nop,ncount,maxiter
     real chimin,temp

     temp=9999.0
     chimin=9999.0


c    ******** INPUT *******
     write(*,*)'HOW MANY LAYERS ?'
     read(*,*)no


c    Read field data from INPUT file
     write(*,*)'ENTER NAME OF INPUT FILE (MAX 20
c    CHARACTERS)>'
     read(*,'(a20)')infile
     write(*,*)'NO. OF DATA POINTS='
     read(*,*)n
     open(unit=21,file=infile,status='unknown')
     read(21,*)(elecsep(i),rapf(i),i=1,n)
     close(21)

     nop=2*no-1


c    Smoothening of INPUT data
     call cubspl(elecsep,rapf,n,raf,ncount,xval)
```

```
      n=ncount
      write(*,*)(xval(i),raf(i),i=1,n)


c     Open OUTPUT files to store data
      write (*,*) 'ENTER NAME OF THE OUTPUT FILE (MAX. 20
c     CHARACTERS)>'
      read(*,'(a20)')outfile
      open(unit=8,file=outfile,status='unknown')
      open(unit=10,file='error.out',status='unknown')
      write(8,*)
      write(8,*)
      x1=xval(1)
      m=n


c     Online User Input : initial choice of layer parameters
      write(*,*)'ENTER INITIAL CHOICE OF LAYER RESISTIVITIES'
      read(*,*)(r(i),i=1,no)
      write(*,*)'ENTER INITIAL CHOICE FOR LAYER THICKNESS'
      read(*,*)(h(i),i=1,(no-1))

      do i=1,no
      b(i)=r(i)
      end do
      do i=1,(no-1)
      b(i+no)=h(i)
      end do
      do i=1,nop
      bini(i)=b(i)
      bak(1,i)=b(i)
      end do


c     Input maximum number of iterations to be performed
      write(*,*)'ENTER MAX NUMBER OF ITERATIONS'
      read(*,*)maxiter

      write(8,*)'INITIAL CHOICE OF MODEL PARAMETERS'
      write(8,*)(b(i),i=1,nop)
      write(8,*)

      write(8,*)'MODIFIED MODEL PARAMETERS'

      do kis=1,maxiter
      chi(kis)=0.0
      e(kis)=0.0
      end do
```

```
kk=0.001
do ii=2,maxiter
do il=1,1200
do ia=1,1200
d(il,ia)=0.0
c(il,ia)=0.0
end do
end do
call appres(no,m,b,tt,x1,0,raf)
do j=1,m
rat(j)=tt(j)
continue
end do
do i=1,m
g(i)=raf(i)-rat(i)
end do
sum=0.0
sum1=0.0
do kd=1,m
sum=sum+g(kd)**2
sum1=sum1+g(kd)**2/raf(kd)
continue
end do
e(ii)=sum
sigma=e(ii)
call chsone(raf,rat,m,1,df(ii),chi(ii),prob(ii))
if(ii.eq.1200)goto 9999
if(ii.gt.2)then
if(chi(ii)-chi(ii-1))11,12,13
11     kk=kk/10.0
goto 37
12     kk=kk
goto 37
13     kk=kk*10.0
goto 37
endif
37     call gena(b,a,x1,nop,m,no,raf)
do i1=1,m
do j1=1,nop
up(i1,j1)=a(i1,j1)
c      if(up(i1,j1).eq.0.0)up(i1,j1)=1.e-20
end do
end do
```

```
call svd(m,nop,1200,1200,up,vp,qp)
wmax=0
do j=1,nop
if(qp(j).gt.wmax)wmax=qp(j)
end do
wmin=wmax*1.0e-5
do j=1,nop
if(qp(j).lt.wmin)qp(j)=0.
end do

do i=1,nop
do j=1,m
ut(i,j)=up(j,i)
end do
do k=1,nop
qi(i)=qp(i)/(qp(i)+kk)**2
vqi(k,i)=vp(k,i)*qi(i)
end do
end do
call inprod(1200,1200,nop,m,ut,g,utd)
call inprod(1200,1200,nop,nop,vqi,utd,mmt)
do il=1,nop
b(il)=b(il)+mmt(il)
if(mmt(il).lt.1.e-6)mmt(il)=1.e-5
if(b(il).lt.0.0)b(il)=abs(b(il))
if(b(il).gt.(bini(il)*10.0))b(il)=bini(il)
bak(ii,il)=b(il)
end do

chi(1)=chimin
if(chimin.gt.chi(ii))then
chimin=chi(ii)
else
temp=chi(ii)
chi(ii)=chimin
end if

write(*,*) 'Iter = ',ii,'      Error = ',chi(ii)
write(10,*)ii,chi(ii)
if(chimin.eq.chi(ii))then
chi(ii)=temp
end if

999   continue
end do
```

```
9999 ino=2
     uin=chi(2)
     chi(1200+2)=1.0e+30
     do klb=3,ii
     if(chi(klb).lt.uin)then
     uin=chi(klb)
     ino=klb
     endif
     end do
     do ii=1,199

     write(8,*)ii,' : ',(bak(ii,ik),ik=1,nop),'  ',chi(ii+1)

     end do

     close(10)

     write(8,*)
     write(8,*)uin,ino,(bak(ino,ik),ik=1,nop)

     write(8,*)
     write(8,*)
     write(8,*)'the minimum chi-sq error is=',uin
     write(8,*)'corr. deg. of freedom is= ',df(ii+1)
     write(8,*)'corr. probability is= ',prob(ii+1)
     write(8,*)
     write(8,*)
     do ik=1,nop
     b(ik)=bak(ino-1,ik)
     end do
     open(unit=9,file="svd.out",status="unknown")
     call appres(no,m,b,tt,x1,1,raf)
     close(9)
     call gena(b,a,x1,nop,m,no,raf)
     call matrans(a,m,nop,bt)
     call matmul(bt,nop,m,a,nop,c)
     call gaussj(c,nop,1200,sol,nop,1200,keep1)
     do kf=1,nop
     do kff=1,nop
     keep2(kf,kff)=sigma*keep1(kf,kff)
     end do
     end do
     do kb=1,nop
     do kj=1,nop
     temp=sqrt(keep2(kb,kb))*sqrt(keep2(kj,kj))
     if(temp.eq.0.0)temp=1.e-20
```

```
      cor(kb,kj)=keep2(kb,kj)/temp
      end do
      end do
      close(8)

      stop
      end

c     Subroutine 'matrans' for matrix transformation
      subroutine matrans(ap,r,c,bp)
      real ap(1200,1200),bp(1200,1200)
      integer r,c
      do i=1,r
      do j=1,c
      bp(j,i)=ap(i,j)
      end do
      end do

      return
      end

      subroutine matmul(au,m,l,bu,n,cu)
      real au(1200,1200),bu(1200,1200),cu(1200,1200)
      integer m,l,n
      do i=1,m
      do j=1,n
      cu(i,j)=0.0
      do ind=1,l
      cu(i,j)=cu(i,j)+au(i,ind)*bu(ind,j)
      end do
      end do
      end do

      return
      end

c     Generate new model
      subroutine gena(bn,an,x1,nop,m,no,raf)
      real bn(1200),an(1200,1200),e,tt(1200)
      real work(1200,6),x1,c1,c2
      integer no,m
      e=0.001
      do i=1,nop
      bn(i)=bn(i)-2.0*e
      call appres(no,m,bn,tt,x1,0,raf)
```

```
      do j=1,m
      work(j,1)=tt(j)
      end do
      bn(i)=bn(i)+e
      call appres(no,m,bn,tt,x1,0,raf)
      do j=1,m
      work(j,2)=tt(j)
      end do
      bn(i)=bn(i)+2.0*e
      call appres(no,m,bn,tt,x1,0,raf)
      do j=1,m
      work(j,3)=tt(j)
      end do
      bn(i)=bn(i)+e
      call appres(no,m,bn,tt,x1,0,raf)
      do j=1,m
      work(j,4)=tt(j)
      end do
      bn(i)=bn(i)-2.0*e
      do j=1,m
      c1=8.0*work(j,3)-work(j,4)
      c2=work(j,1)-8.0*work(j,2)
      an(j,i)=(c1+c2)/(12.0*e)
      end do
      end do

      return
      end


c     Calculate the inter prod distance
      subroutine inprod(mm,nn,m,n,amx,v1,v2)
      dimension v1(nn),v2(mm),amx(mm,nn)
      do i=1,m
      sum=0.0
      do j=1,n
      sum=v1(j)*amx(i,j)+sum
      end do
      v2(i)=sum
      end do

      return
      end

c     Subroutine Singular Value Decomposition
```

```
subroutine svd(m,n,mp,np,uk,vk,w)
integer m,mp,n,np,nmax
real w(np),vk(np,np),uk(mp,np)
parameter (nmax=500)
integer i,its,j,jj,k,l,nm
real anorm,c,f,g,h,s,scale,x,y,z,rvi(nmax),pythag
g=0.0
scale=0.0
anorm=0.0
do i=1,n
l=i+1
rvi(i)=scale*g
g=0.0
s=0.0
scale=0.0
if(i.le.m)then
do k=i,m
scale=scale+abs(uk(k,i))
end do
if(scale.ne.0.0)then
do k=i,m
uk(k,i)=uk(k,i)/scale
s=s+uk(k,i)*uk(k,i)
end do
f=uk(i,i)
g=-sign(sqrt(s),f)
h=f*g-s
uk(i,i)=f-g
do j=l,n
s=0.0
do k=i,m
s=s+uk(k,i)*uk(k,j)
end do
c        if(h.eq.0.0)h=1.e-20
f=s/h
do k=i,m
uk(k,j)=uk(k,j)+f*uk(k,i)
end do
end do
end do
do k=i,m
uk(k,i)=scale*uk(k,i)
end do
end if
```

```
        endif
        w(i)=scale*g
        g=0.0
        s=0.0
        scale=0.0
        if((i.le.m).and.(i.ne.n))then
        do k=l,n
        scale=scale+abs(uk(i,k))
        end do
        if (scale.ne.0.0)then
        do k=l,n
        uk(i,k)=uk(i,k)/scale
        s=s+uk(i,k)*uk(i,k)
        end do
        f=uk(i,l)
        g=-sign(sqrt(s),f)
        h=f*g-s
        uk(i,l)=f-g
c       if(h.eq.0.0)h=1.e-20
        do k=l,n
        rvi(k)=uk(i,k)/h
        end do
        do j=l,m
        ɛ=0.0
        do k=l,n
        s=s+uk(j,k)*uk(i,k)
        end do
        do k=l,n
        uk(j,k)=uk(j,k)+s*rvi(k)
        end do
        end do
        do k=l,n
        uk(i,k)=scale*uk(i,k)
        end do
        end if
        end if
        anorm=max(anorm,(abs(w(i))+abs(rvi(i))))
        end do
        do i=n,1,-1
        if(i.lt.n)then
        if(g.ne.0.0)then
        do j=l,n
c       if(uk(i,l).eq.0.0)uk(i,l)=1.e-20
```

```
vk(j,i)=(uk(i,j)/uk(i,l))/g
end do
do j=l,n
s=0.0
do k=l,n
s=s+uk(i,k)*vk(k,j)
end do
do k=l,n
vk(k,j)=vk(k,j)+s*vk(k,i)
end do
end do
end if
do j=l,n
vk(i,j)=0.0
vk(j,i)=0.0
end do
end if
vk(i,i)=1.0
g=rvi(i)
l=i
end do
do i=min(m,n),1,-1
l=i+1
g=w(i)
do j=l,n
uk(i,j)=0.0
end do
if(g.ne.0.0)then
g=1.0/g
do j=l,n
s=0.0
do k=l,m
s=s+uk(k,i)*uk(k,j)
end do
c      if(uk(i,i).eq.0.0)uk(i,i)=1.e-20
f=(s/uk(i,i))*g
do k=i,m
uk(k,j)=uk(k,j)+f*uk(k,i)
end do
end do
do j=i,m
uk(j,i)=uk(j,i)*g
end do
```

```
        else
        do j=i,m
        uk(j,i)=0.0
        end do
        end if
        uk(i,i)=uk(i,i)+1.0
        end do
        do k=n,1,-1
        do its=1,30
        do l=k,1,-1
        nm=l-1
        if((abs(rvi(l))+anorm).eq.anorm) goto 2
        if((abs(w(nm))+anorm).eq.anorm) goto 1
        end do
1       c=0.0
        s=1.0
        do i=l,k
        f=s*rvi(i)
        rvi(i)=c*rvi(i)
        if((abs(f)+anorm).eq.anorm) goto 2
        g=w(i)
        h=pythag(f,g)
        w(i)=h
c       if(h.eq.0.0)h=1.e-20
        h=1.0/h
        c=(g*h)
        s=-(f*h)
        do j=1,m
        y=uk(j,nm)
        z=uk(j,i)
        uk(j,nm)=(y*c)+(z*s)
        uk(j,i)=-(y*s)+(z*c)
        end do
        end do
2       z=w(k)
        if(l.eq.k)then
        if(z.lt.0.0)then
        w(k)=-z
        do j=1,n
        vk(j,k)=-vk(j,k)
        end do
        end if
        goto 3
```

```
      end if
      if(its.eq.30) pause 'no convergence in svd'
      x=w(l)
      nm=k-1
c     if(w(nm).eq.0.0)w(nm)=1.e-20
c     if(rvi(k).eq.0.0)rvi(k)=1.e-20
      y=w(nm)
      g=rvi(nm)
      h=rvi(k)
      f=((y-z)*(y+z)+(g-h)*(g+h))/(2.0*h*y)
      g=pythag(f,1.0)
c     if(x.eq.0.0)x=1.e-20
c     if(f.eq.0.0)f=1.e-20
      f=((x-z)*(x+z)+h*((y/(f+sign(g,f)))-h))/x
      c=1.0
      s=1.0
      do j=1,nm
      i=j+1
      g=rvi(i)
      y=w(i)
      h=s*g
      g=c*g
      z=pythag(f,h)
      rvi(j)=z
c     if(z.eq.0.0)z=1.e-20
      c=f/z
      s=h/z
      f=(x*c)+(g*s)
      g=-(x*s)+(g*c)
      h=y*s
      y=y*c
      do jj=1,n
      x=vk(jj,j)
      z=vk(jj,i)
      vk(jj,j)=(x*c)+(z*s)
      vk(jj,i)=-(x*s)+(z*c)
      end do
      z=pythag(f,h)
      w(j)=z
      if(z.ne.0.0)then
      z=1.0/z
      c=f*z
      s=h*z
```

```
        end if
        f=(c*g)+(s*y)
        x=-(s*g)+(c*y)
        do jj=1,m
        y=uk(jj,j)
        z=uk(jj,i)
        uk(jj,j)=(y*c)+(z*s)
        uk(jj,i)=-(y*s)+(z*c)
        end do
        end do
        rvi(l)=0.0
        rvi(k)=f
        w(k)=x
        end do
3       continue
        end do

        return
        end

c       Function 'pythg' to calculate the Pythagorus Equation
        function pythag(awr,bwr)
        real awr,bwr,pythag
        real absa,absb
        absa=abs(awr)
        absb=abs(bwr)
        if(absa.gt.absb)then
        if(absa.eq.0.0)absa=1.e-20
        pythag=absa*sqrt(1.+(absb/absa)**2)
        else
        if(absb.eq.0.)then
        pythag=1.e-20
        else
        pythag=absb*sqrt(1.+(absa/absb)**2)
        end if
        end if

        return
        end

        subroutine gaussj(ka,n,np,gb,m,mp,ga)
        integer m,mp,n,np,nmax
        real ga(np,np),gb(np,np),ka(np,np)
        parameter (nmax=50)
```

```
      integer i,icol,irow,j,k,l,ll,indxc(nmax),indxr(nmax),
*     ipiv(nmax)
      real big,dum,pivinv
      m=n
      do i=1,np
      do j=1,np
      ga(i,j)=ka(i,j)
      end do
      end do
      do j=1,n
      ipiv(j)=0
      end do
      do i=1,n
      big=0
      do j=1,n
      if(ipiv(j).ne.1)then
      do k=1,n
      if(ipiv(k).eq.0)then
      if(abs(ga(j,k)).ge.big)then
      big=abs(ga(j,k))
      irow=j
      icol=k
      end if
      else if(ipiv(k).gt.1)then
      pause 'singular matrix in gaussj'
      end if
      end do
      end if
      end do
      ipiv(icol)=ipiv(icol)+1
      if(irow.ne.icol)then
      do l=1,n
      dum=ga(irow,l)
      ga(irow,l)=ga(icol,l)
      ga(icol,l)=dum
      end do
      do l=1,m
      dum=gb(irow,l)
      gb(irow,l)=gb(icol,l)
      gb(icol,l)=dum
      end do
      end if
      indxr(i)=irow
```

```
indxc(i)=icol
if(ga(icol,icol).eq.0.)pause 'singular matrix in gaussj'
pivinv=1./ga(icol,icol)
ga(icol,icol)=1
do l=1,n
ga(icol,l)=ga(icol,l)*pivinv
end do
do l=1,m
gb(icol,l)=gb(icol,l)*pivinv
end do
do ll=1,n
if(ll.ne.icol)then
dum=ga(ll,icol)
ga(ll,icol)=0
do l=1,n
ga(ll,l)=ga(ll,l)-ga(icol,l)*dum
end do
do l=1,m
gb(ll,l)=gb(ll,l)-gb(icol,l)*dum
end do
end if
end do
end do
do l=n,1,-1
if(indxr(l).ne.indxc(l))then
do k=1,n
dum=ga(k,indxr(l))
ga(k,indxr(l))=ga(k,indxc(l))
ga(k,indxc(l))=dum
end do
end if
end do

return
end

subroutine chsone(bins,ebins,nbins,knstrn,df,chisq,prob)
integer knstrn,nbins
real chisq,df,prob,bins(nbins),ebins(nbins)
integer j
real gammq
df=nbins-knstrn
chisq=0.
do j=1,nbins
```

```
if(ebins(j).le.0.)pause 'bad expected no. in chsone'
chisq=chisq+(bins(j)-ebins(j))**2/ebins(j)
end do
chisq=chisq/nbins
prob=gammq(0.5*df,0.5*chisq)

return
end


function gammq(ra,ax)
real ra,ax,gammq
real gammcf,gamscr,gln
if(ax.lt.0..or.ra.le.0.)pause 'bad argument in gammq'
if(ax.lt.ra+1.)then
call gser(gamscr,ra,ax,gln)
gammq=1.-gamscr
else
call gcf(gammcf,ra,ax,gln)
gammq=gammcf
end if

return
end


subroutine gser(gamser,ra,ax,gln)
integer itmax
real ra,gamser,gln,ax,eps
parameter (itmax=100,eps=3.e-7)
integer n
real apc,del,sum,gammln
gln=gammln(ra)
if(ax.le.0.)then
if(ax.lt.0.)pause 'ax < 0 in gser'
gamser=0
return
end if
apc=ra
sum=1./ra
del=sum
do n=1,itmax
apc=apc+1
del=del*ax/apc
sum=sum+del
if(abs(del).lt.abs(sum)*eps)goto 1
```

```
      end do
      pause'ra too large,itmax too small in gser'
1     gamser=sum*exp(-ax+ra*log(ax)-gln)

      return
      end


      subroutine gcf(gammcf,ra,ax,gln)
      integer itmax
      real ra,gammcf,gln,ax,eps,fpmin
      parameter (itmax=100,eps=3.e-7,fpmin=1.e-30)
      integer i
      real can,chb,chd,cdel,ch,gammln,chc
      gln=gammln(ra)
      chb=ax+1.-ra
      chc=1./fpmin
      chd=1./chb
      ch=chd
      do i=1,itmax
      can=-i*(i-ra)
      chb=chb+2
      chd=can*chd+chb
      if(abs(chd).lt.fpmin)chd=fpmin
      chc=chb+can/chc
      if(abs(chc).lt.fpmin)chc=fpmin
      chd=1./chd
      cdel=chd*chc
      ch=ch*cdel
      if(abs(cdel-1.).lt.eps)goto 1
      end do
      pause 'a too large,itmax too small in gcf'
1     gammcf=exp(-ax+ra*log(ax)-gln)*ch

      return
      end


      function gammln(xx)
      real gammln,xx
      integer j
      double precision ser,stp,tmp,x,y,cof(6)
      save cof,stp
      data cof,stp/76.18009172947146d0,
*     -86.50532032941677d0,
*     24.01409824083091d0,-1.231739572450155d0,
```

```
*       .1208650973866179d-2,
*       -.5395239384953d-5,2.5066282746310005d0/
        x=xx
        y=x
        tmp=x+5.5d0
        tmp=(x+0.5d0)*log(tmp)-tmp
        ser=1.000000000190015d0
        do j=1,6
        y=y+1.d0
        ser=ser+cof(j)/y
        end do
        gammln=tmp+log(stp*ser/x)

        return
        end


c       Run forward program to generate App.Res. vs. AB/2 data
        subroutine appres(i9,n,bb,ss,x1,in,raf)
        real bb(1200),r(50),d(50),t(35),ss(1200),f,y,x1
        real u,b,a1,a2,axis(1200),raf(1200),ssg(1200)
        integer i9,i8,n,m,in
        x1=1
        do i=1,i9
        r(i)=bb(i)
        d(i)=bb(i+i9)
        end do
        f=exp(alog(10.0)/8)
        i8=i9-1
        y=x1/822.8
        do j=1,34
        b=r(i9)
        do k=1,i8
        i=i9-k
        u=d(i)/y
        if(5.0-u)14,14,15
14      b=r(i)
        goto 13
15      a1=exp(u)
        a2=(a1-1.0/a1)/(a1+1.0/a1)
        b=(b+a2*r(i))/(1.0+a2*b/r(i))
13      end do
        t(j)=b
        y=y*f
        end do
```

```
     do ks=1,1200
     ss(ks)=0.0
     end do
     do m=1,n
     b=r(i9)
     do k=1,i8
     i=i9-k
     u=d(i)/y
     if(5.0-u)46,46,45
46   b=r(i)
     goto 44
45   a1=exp(u)
     a2=(a1-1.0/a1)/(a1+1.0/a1)
     b=(b+a2*r(i))/(1.0+a2*b/r(i))
44   end do
     t(35)=b
     y=y*f
     s=0.0
     s=42.*t(1)-103.*t(3)+144.*t(5)-211.*t(7)+330.*t(9)-574.*t(11)
     s=s+1184.*t(13)-3162.*t(15)+10219.*t(17)-24514.*t(19)+18192.*t(21)
     s=s+6486.*t(23)+1739.*t(25)+79.*t(27)+1200.*t(29)-106.*t(31)
     s=s+93.*t(33)-38.*t(35)
     s=s/10000.0
     ss(m)=s
     do j=1,34
     t(j)=t(j+1)
     end do
     end do
     if(n.lt.8)then
     do m=1,n
     axis(m)=exp((m*log(10.))/8)
     end do
     else
     do m=1,8
     axis(m)=exp((m*log(10.))/8)
     end do
     if(n.lt.16)then
     do m=9,n
     axis(m)=axis(m-8)*10
     end do
     else
     do m=9,16
     axis(m)=axis(m-8)*10
```

```
end do
if(n.lt.24)then
do m=17,n
axis(m)=axis(m-16)*100
end do
else
do m=17,24
axis(m)=axis(m-16)*100
end do
if(n.lt.32)then
do m=25,n
axis(m)=axis(m-24)*1000
end do
else
do m=25,32
axis(m)=axis(m-24)*1000
end do
if(n.gt.32)then
do m=33,n
axis(m)=axis(m-32)*10000
end do
end if
end if
end if
end if
end if
if(in.eq.1)then
do m=1,n
write(9,*)axis(m),ss(m)
end do
do m=1,n
ssg(m)=alog10(ss(m))
end do
end if

return
end
```

```
c     Interpolate the input data using cubic spline method
      subroutine cubspl(elecsep,rapf,nres,rappf,ncount,xval)
      real elecsep(1200),rapf(1200),rappf(1200),xval(1200)
      integer nres,ncount,m
      real yp1,ypn,y2(1200)
      yp1=(rapf(2)-rapf(1))/(elecsep(2)-elecsep(1))
```

```
ypn=(rapf(nres)-rapf(nres-1))/(elecsep(nres)-elecsep(nres-1))
if(nres.lt.8)then
do m=1,nres
xval(m)=exp((m*log(10.))/8)
end do
else
do m=1,8
xval(m)=exp((m*log(10.))/8)
end do
if(nres.lt.16)then
do m=9,nres
xval(m)=xval(m-8)*10
end do
else
do m=9,16
xval(m)=xval(m-8)*10
end do
if(nres.lt.24)then
do m=17,nres
xval(m)=xval(m-16)*100
end do
else
do m=17,24
xval(m)=xval(m-16)*100
end do
if(nres.lt.32)then
do m=25,nres
xval(m)=xval(m-24)*1000
end do
else
do m=25,32
xval(m)=xval(m-24)*1000
end do
f(nres.gt.32)then
do m=33,nres
xval(m)=xval(m-32)*10000
end do
end if
end if
end if
end if
end if
call spline(elecsep,rapf,nres,yp1,ypn,y2)
```

```fortran
      do i=1,nres
      call splint(elecsep,rapf,y2,nres,xval(i),rappf(i))
      end do
      ncount=nres

      return
      end


c     Subroutine 'spline' to interpolate the input data by spline method
      subroutine spline(elecsep,rapf,nres,yp1,ypn,y2)
      integer nres,nmax
      real yp1,ypn,elecsep(1200),rapf(1200),y2(1200)
      parameter (nmax=500)
      integer i,k
      real p,qn,sig,un,u(nmax)

      if(yp1.gt.0.99e30)then
      y2(1)=0.
      u(1)=0.
      else
      y2(1)=-0.5
      u(1)=(3./(elecsep(2)-elecsep(1)))*((rapf(2)-rapf(1))/
     *(elecsep(2)-elecsep(1))-yp1)
      end if
      do i=2,nres-1
      sig=(elecsep(i)-elecsep(i-1))/(elecsep(i+1)-elecsep(i-1))
      p=sig*y2(i-1)+2.
      y2(i)=(sig-1.0)
      u(i)=(rapf(i+1)-rapf(i))/(elecsep(i+1)-elecsep(i))-
     *(rapf(i)-rapf(i-1))/(elecsep(i)-elecsep(i-1))
      u(i)=(6.0*u(i)/(elecsep(i+1)-elecsep(i-1))-sig*u(i-1))/p
      end do
      if(ypn.gt.0.99e30)then
      qn=0.
      un=0.
      else
      qn=0.5
      un=(3./(elecsep(nres)-elecsep(nres-1)))*
     *(ypn-(rapf(nres)-rapf(nres-1))/(elecsep(nres)-elecsep(nres-1)))
      end if
      y2(nres)=(un-qn*u(nres-1))/(qn*y2(nres-1)+1.)
      do k=nres-1,1,-1
      y2(k)=y2(k)*y2(k+1)+u(k)
      end do
```

```
      return
      end

      subroutine splint(xa,ya,y2a,n,x,y)
      integer n
      real x,y,xa(*),y2a(*),ya(*)
      integer k,khi,klo
      real a,b,h
      klo=1
      khi=n
1     if(khi-klo.gt.1)then
      k=(khi+klo)/2
      if(xa(k).gt.x)then
      khi=k
      else
      klo=k
      end if
      goto 1
      end if
      h=xa(khi)-xa(klo)
      if(h.eq.0.)pause'bad inpit into splint'
      a=(xa(khi)-x)/h
      b=(x-xa(klo))/h
      y=a*ya(klo)+b*ya(khi)+
*     ((a**3-a)*y2a(klo)+(b**3-b)*y2a(khi)*(h**2))/6

      return
      end
```

Guidelines for Operating the Inversion Program Using Evolutionary Programming Algorithm Given in Appendix 4.3

Here, upper and lower limits of layer parameters and apparent resistivity ($\bar{\rho}$) versus half electrode spacing ($AB/2$) data are given as INPUT and final layer parameters and corresponding chi-square error are obtained as OUTPUT.

INPUT data required :

Layer parameters :
(i) Number of layers
(ii) Upper and lower limits of layer resistivities
(iii) Upper and lower limits of layer thicknesses

Feed the layer parameters in the following sequence :
(i) Top layer parameters
(ii) Then, going down step by step.

Other data :
(i) Name of INPUT field acquired data file ($\bar{\rho}$ vs. $AB/2$)
(ii) Number of data points in the INPUT file
(iii) Random seed number
(iv) Number of iteration

OUTPUT

The final layer parameters and chi-square error are obtained as output. The result is stored in a file named 'GA.out' and the error after each iteration is stored in a file called 'error.out'.

SAMPLE INPUT:
Number of layers = 3
Model Parameter data limits = 150.0 300.0
    (Lower & Upper)

20.0 40.0	resistivity
40.0 60.0	
2.1 2.8	
9.8 13.3	thickness

Input field data file Name : field.dat*
Number of data points = 19
Randon seed number = 2
Number of iteration = 1000
[* Sample field data stored in File "field.dat" : (H-Type curve)

AB/2	$\bar{\rho}$
1.5	9.0
2.0	7.7
3.0	6.95
4.0	6.9
5.0	7.2
6.0	7.5
8.0	8.3
10.0	8.8
15.0	10.5
20.0	12.1
30.0	15.1
40.0	17.5
50.0	19.1
60.0	21.0
80.0	23.0
100.0	24.6
120.0	25.3
150.0	25.8
200.0	25.8]

SAMPLE OUTPUT:

Resistivity of Layer - 1	= 61.84
Resistivity of Layer - 2	= 7.37
Resistivity of Layer - 3	= 29.60
Thickness of Layer - 1	= 0.22
Thickness of Layer - 2	= 6.23

Chi-square Error = 5.92E-02

Appendix 4.3

c Inversion of Apparent Resistivity Data Using Evolutionary Programming Technique

```
c      Global variable declaration
       integer maxiter,gener
       real param,fittest(100)
       real rapf(50),elecsep(50)
       integer no_layer,no_param,i,j,nopoint,seed
       real low_lmt(50),up_lmt(50)

       common/val/bak_parent(50,100),bak_point(50,100)
       common/val1/parent(50,100),chi(100)

c      set parameter to minimum chi-square value expected
       param=.01

c      Open input and output files.
       call opn_file

c      Online user input data
       call getdat(no_layer,no_param,low_lmt,up_lmt,nopoint,elecsep,
*      rapf,seed,maxiter)

c      Generation of 50 models using lower limit and upper limit of layer
c      parameters
       call gen_pop(no_param,low_lmt,up_lmt,seed)

c      Run forward program to generate App.Res data vs. AB/2 for each
c      model
       call appres(no_layer,nopoint,1)

c      calculate chi-square error of each model
       call chi_err(rapf,nopoint,no_param,1)

       do gener=1,maxiter

c      Set the first model as the fittest model
       do i=1,no_param
       fittest(i)=parent(i,1)
       end do
```

```
c       Call subroutine to create n set of mutated models from n set of
c       models
        call mutation(no_param,up_lmt,low_lmt,seed)

c       add the new n models with previous n models to create 2n models
        do j=1,50
        do i=1,no_param
        parent(i,j+50)=bak_parent(i,j)
        end do
        end do

c       Calling subroutine 'appres' to generate App.Res vs. AB/2 data for
c       new n models
        call appres(no_layer,nopoint,51)

c       call chi_err to calculate new chi value for each 2n model
        call chi_err(rapf,nopoint,no_param,51)

c       Sort the models according to their fitness
        call fit_sort(no_param)

c       Set the model with least chi-square error as the fittest model
        do i=1,no_param
        fittest(i)=parent(i,1)
        end do

c       Print iteration and least chi-square error
        write(*,*)gener,' chi ',chi(1)

c       Check the termination criteria and print result
        if(chi(1).le.param.or.gener.eq.maxiter)then
        call result(chi(1),fittest,no_param,no_layer)

        go to 9999
        end if

c       Write iteration and chi-square error in the file 'error.out'
        write(10,*)gener,'',chi(1)
        end do

c       Call subroutine to close all files before termination 9999
        call cls_file

        stop
        end
```

```
c     Subroutine 'mutation' creates n number of mutated models from
c     n number of input models
      subroutine mutation(no_param,up_lmt,low_lmt,seed)

      real mut_par(50,100),up_lmt(50),low_lmt(50)
      real mutval,rand
      integer seed

      common/val/bak_parent(50,100),bak_point(50,100)
      common/val1/parent(50,100),chi(100)

      do j=1,50
      call stepfunc(chi(j),mutval,seed)
      do i=1,no_param
      rand=ran(seed)
      if(rand.gt.0.5)then
      mut_par(i,j)=parent(i,j)*mutval
      else
      mut_par(i,j)=parent(i,j)/mutval
      end if
      end do
      do i=1,no_param
      bak_parent(i,j)=mut_par(i,j)
      end do
      end do
      return
      end


c     Subroutine 'chi_err' calculates chi-square error for each model
      subroutine chi_err(rapf,nopoint,no_param,iter)
      real rapf(50)
      integer i,nopoint,no_param,iter

      common/val/bak_parent(50,100),bak_point(50,100)
      common/val1/parent(50,100),chi(100)

      do j=iter,iter+49
      do i=1,nopoint
      chi(j)=chi(j)+abs(((rapf(i)-bak_point(i,j))**2)/
      bak_point(i,j))
      end do
      chi(j)=chi(j)/nopoint
      end do

      return
      end
```

```
c    Subroutine 'result' prints the layer parameters and chi-square error
c    of the converged model
     subroutine result(chi,fittest,no_param,no_layer)

     real fittest(50)
     integer no_param,no_layer

     write(9,*)'###### OUTPUT OF INVERSION PROGRAM ######'
     write(*,*)'THE MODEL HAS CONVERGED'
     write(*,*)'The MODEL PARAMETERS ARE ::'
     write(9,*)'THE MODEL HAS CONVERGED'
     write(9,*)'The MODEL PARAMETERS ARE ::'
     do i=1,no_layer
     write(9,*)'Resistivity of Layer - ',i,' = ',fittest(i)
     write(*,*)'Resistivity of Layer - ',i,' = ',fittest(i)
     end do
     do i=no_layer+1,no_param
     write(9,*)'Thickness of Layer - ',i-no_layer,' = ',fittest(i)
     write(*,*)'Thickness of Layer - ',i-no_layer,' = ',fittest(i)
     end do
     write(9,*)'CHI_SQUARE ERROR :: ',chi
     write(*,*)'CHI_SQUARE ERROR :: ',chi

     return
     end


c    Function 'ran' generates random number using the seed value
     function ran(idum)

     integer idum,im1,im2,imm1,ia1,
*    ia2,iq1,iq2,ir1,ir2,ntab,ndiv
     real ran,am,eps,rnmax
     parameter (im1=2147483563,im2=2147483399,
*    am=1./im1,imm1=im1-1,ia1=40014,ia2=40692,
*    iq1=53668,iq2=52774,ir1=12211,ir2=3791,
*    ntab=32,ndiv=1+imm1/ntab,eps=1.2e-7,
*    rnmax=1.-eps)
     integer idum2,j,k,iv(ntab),iy
     save iv,iy,idum2
     data idum2/123456789/,iv/ntab*0/,iy/0/

     if(idum.le.0.0)then
     idum=max(-idum,1)
     idum2=idum
     do j=ntab+8,1,-1
     k=idum/iq1
```

```
idum=ia1*(idum-k*iq1)-k*ir1
if(idum.lt.0)idum=idum+im1
if(j.le.ntab)iv(j)=idum
end do
iy=iv(1)
end if
k=idum/iq1
idum=ia1*(idum-k*iq1)-k*ir1
if(idum.lt.0)idum=idum+im1
k=idum2/iq2
idum2=ia2*(idum2-k*iq2)-k*ir2
if(idum2.lt.0)idum2=idum2+im2
j=1+iy/ndiv
iy=iv(j)-idum2
iv(j)=idum
if(iy.lt.1)iy=iy+imm1
ran=min(am*iy,rnmax)

return
end


c    Subroutine 'appres is a forward program which generates App.Res
c    vs. AB/2
c    data for each model
     subroutine appres(i9,n,init)

     real r(50),d(50),t(35),f,y,x1
     real u,b,a1,a2,axis(50)
     integer i9,i8,n,m,new,init

     common/val/bak_parent(50,100),ss(50,100)
     common/val1/bb(50,100),chi(100)

     x1=1.0

     do new=init,init+49
     do i=1,i9
     r(i)=bb(i,new)
     d(i)=bb(i+i9,new)
     end do
     f=exp(alog(10.0)/8)
     i8=i9-1
     y=x1/822.8
     do j=1,34
     b=r(i9)
     do k=1,i8
```

```
         i=i9-k
         u=d(i)/y
         if(5.0-u)14,14,15
14       b=r(i)
         goto 13
15       a1=exp(u)
         a2=(a1-1.0/a1)/(a1+1.0/a1)
         b=(b+a2*r(i))/(1.0+a2*b/r(i))
13       end do
         t(j)=b
         y=y*f
         end do
         do ks=1,50
         ss(ks,new)=0.0
         end do
         do m=1,n
         b=r(i9)
         do k=1,i8
         i=i9-k
         u=d(i)/y
         if(5.0-u)46,46,45
46       b=r(i)
         goto 44
45       a1=exp(u)
         a2=(a1-1.0/a1)/(a1+1.0/a1)
         b=(b+a2*r(i))/(1.0+a2*b/r(i))
44       end do
         t(35)=b
         y=y*f
         s=0.0
         s=42.*t(1)-103.*t(3)+144.*t(5)-211.*t(7)+330.*t(9)-574.*t(11)
         s=s+1184.*t(13)-3162.*t(15)+10219.*t(17)-24514.*t(19)+18192.*t(21)
         s=s+6486.*t(23)+1739.*t(25)+79.*t(27)+200.*t(29)-106.*t(31)
         s=s+93.*t(33)-38.*t(35)
         s=s/10000.0
         ss(m,new)=s
         do j=1,34
         t(j)=t(j+1)
         end do
         end do
         if(n.lt.8)then
         do m=1,n
         axis(m)=exp((m*log(10.))/8)
```

```
      end do
      else
      do m=1,8
      axis(m)=exp((m*log(10.))/8)
      end do
      if(n.lt.16)then
      do m=9,n
      axis(m)=axis(m-8)*10
      end do
      else
      do m=9,16
      axis(m)=axis(m-8)*10
      end do
      if(n.lt.24)then
      do m=17,n
      axis(m)=axis(m-16)*100
      end do
      else
      do m=17,24
      axis(m)=axis(m-16)*100
      end do
      if(n.lt.32)then
      do m=25,n
      axis(m)=axis(m-24)*1000
      end do
      else
      do m=25,32
      axis(m)=axis(m-24)*1000
      end do
      if(n.gt.32)then
      do m=33,n
      axis(m)=axis(m-32)*10000
      end do
      end if
      end if
      end if
      end if
      end if
      end do

      return
      end


      subroutine getdat(no_layer,no_param,low_lmt,up_lmt,nopoint,
```

```
*       elecsep,rapf,seed,maxiter)
        integer no_layer,no_param,nopoint,seed,maxiter
        real low_lmt(50),up_lmt(50),elecsep(50),rapf(50)
        character*20 infile

        write(*,*)'Enter no. of Layers :'
        read(*,*)no_layer
        write(*,*)'No. of Layers : ',no_layer
        write(9,*)'No.of layers',no_layer
        no_param=2*no_layer-1
        write(*,*)'Enter Model Param. data limits( Lr. - Up.):'
        do i=1,no_param
        read(*,*)low_lmt(i),up_lmt(i)
        write(*,*)low_lmt(i),up_lmt(i)
        end do
        write(9,*)'Initial model parameter limits (User Given)'
        write(9,*)(low_lmt(i),up_lmt(i),i=1,no_param)
        write(*,*)'Enter input field aquired data file name :'
        read(*,'(a20)')infile
        write(*,*)'Enter no. of data points'
        read(*,*)nopoint
        open(unit=8,file=infile,status='old')
        write(9,*)'No. of data points in input file:: ',nopoint
        do i=1,nopoint
        read(8,*)elecsep(i),rapf(i)
        write(9,*)elecsep(i),rapf(i)
        end do
        write(*,*)'enter random no. seed'
        read(*,*)seed
        write(9,*)'Random no. seed used ::',seed
        write(*,*)'Enter no. of Iteration'
        read(*,*)maxiter
        write(9,*)'Iteration no. used :',maxiter

        return
        end

c       Subroutine 'stepfunc' generates a mutation value for each model
c       according to their chi-square error
        subroutine stepfunc(error,mutval,seed)

        integer seed
        real error,mutval

        if(error.gt.200.0)mutval=ran(seed)
```

```
if(error.gt.100.0.and.error.le.200.0)mutval=.5
if(error.gt.50.0.and.error.le.100.0)mutval=.7
if(error.gt.25.0.and.error.le.50.0)mutval=.8
if(error.gt.10.0.and.error.le.25.0)mutval=.85
if(error.gt.5.0.and.error.le.10.0)mutval=.9
if(error.gt.2.0.and.error.le.5.0)mutval=.94
if(error.gt.1.5.and.error.le.2.0)mutval=.97
if(error.gt.1.0.and.error.le.1.5)mutval=.98
if(error.gt.0.5.and.error.le.1.0)mutval=.99
if(error.le.0.5)mutval=.9999

return
end


c    Subroutine 'fit_sort' sorts the models according to their chi-square
c    error
     subroutine fit_sort(no_param)

     real tempchi,temp(50)
     integer i,j,m,no_param

     common/val/bak_parent(50,100),bak_point(50,100)
     common/val1/parent(50,100),chi(100)

     do j=1,99
     do i=1,100-j
     if(chi(i).gt.chi(i+1))then
     tempchi=chi(i)
     do m=1,no_param
     temp(m)=parent(m,i)
     end do
     chi(i)=chi(i+1)
     do m=1,no_param
     parent(m,i)=parent(m,i+1)
     end do
     chi(i+1)=tempchi
     do m=1,no_param
     parent(m,i+1)=temp(m)
     end do
     end if
     end do
     end do

     return
     end
```

```
c    Subroutine 'gen_pop' generates n number of initial models
     subroutine gen_pop(no_param,low_lmt,up_lmt,seed)

     real low_lmt(50),up_lmt(50)
     integer i,j,no_param,seed

     common/val1/parent(50,100),chi(100)

     do j=1,50
     do i=1,no_param
     parent(i,j)=abs(((up_lmt(i)-low_lmt(i))*ran(seed))+low_lmt(i))
     end do
     end do

     return
     end


c    Subroutine 'opn_file' opens input and output files
     subroutine opn_file

     open(unit=9,file='GA.out',status='unknown')
     open(unit=10,file='error.out',status='unknown')

     end

c    Subroutine 'cls_file' closes all files before the termination of program
     subroutine cls_file

     close(8)
     close(9)
     close(10)

     end
```